THE
MARKET
ENTRY
STRATEGY

Charting Your Course to
Early Stage Venture Success

JACQUES BENKOSKI

BLUE SAILS PUBLISHING

Building a company is
so much like racing a sailboat

"You need a plan and then the ability to reevaluate the plan in real time as new information is acquired. Suddenly, the wind changes or competitors are stronger than you thought they'd be. Each mark in the race is a milestone where you need to look up and evaluate what you're doing, how you're doing. When you skipper a sailboat, you need to be in the moment. You need to let go of your self-doubt, and yet you can't lose the ability to question your decisions. Most of all, when it's time to make a decision, as competitors come upon you or the wind shifts, you make it. The race won't wait for you. And finally, when you race a sailboat, the selection of your crew is just completely paramount. It's impossible to be an effective skipper if you don't have the right people working harmoniously in the right roles."

—DIANE GREENE, VMware cofounder

Contents

Introduction

As a competitive sailor, I always find inspiration in the analogy that racing a sailboat is like running a startup, as Diane Greene describes in the foreword. One of the most fascinating points to observe is when fledgling startups move from the idealized and optimistic dream of their founders to the reality of what it takes to become a real and enduring business. This transition is often compared to a Darwinian process where only the fittest survive. As a matter of fact, the economic theory underlying venture capital (VC) is the financing of a succession of experiments where the execution of a funding round's objectives opens the road to the next round and—at least theoretically—failure leads to the end of the experiment. Although external factors often play a role, it's interesting to note that business failures happen regardless of financial market conditions, funding accessibility, quality of investors, technology advantage, or target markets. At the root of every business failure, one almost always finds execution missteps, as well as a lack of both alignment and team focus. Of course, at the end of the day (or the runway), every startup fails because it's unable to raise more money but when people decipher the reasons behind why businesses fail, they come up with a long list of culprits, some of which are listed below:

- No need for the product

- Not able to penetrate the market

- Not selling effectively

- Not understanding the competition

- Not having a reliably good product

- Not being effective at marketing

- Having the wrong team

These are some common reasons that companies fail—there are doubtless many others—but as a startup you might wonder, "Is there something fundamental that impacts startup success? Or is it mostly luck or timing?"

As you review the list of reasons why companies fail, are they actually causes or symptoms? Most of these are in fact symptoms because if you ask, "Why wasn't there a need for your product? Why did you run out of money? Why didn't you understand the competition?" then you will get closer to the same root causes listed above. As I reflected on the many modes of failure of startups and on the few that executed seemingly flawlessly, there seem to be common traits and execution models that come to bear when they approach the crucial Darwinian selection of the initial market penetration. It is also clear that the way a company executes this *market entry* stage impacts its trajectory in quite a permanent way for the rest of its life.

Over the years I have been involved with many startups in nearly every capacity. Having started as an executive in a large corporation, I had the opportunity to join a startup at the ground level and experienced the thrill of fast growth, and the joys and challenges of going public and operating as a public company. Running a company as CEO with rapid growth fueled by capital availability, I also dealt with the crash and its aftermath and had to find creative ways to survive and a path to relative success. My journey continued as an executive chairman of both successful and unsuccessful startups, and an advisor and sounding board to many CEOs. This brought the experience of the highs and lows that come from acquiring companies and from being acquired. Along the way came how to run a large business, how to craft multimillion-dollar contracts, and how to execute the painful processes of layoffs, fire sales, and shutdowns. For the last fifteen years I've been a venture capital investor and board member of dozens of companies. Throughout my career, I have made more mistakes than good calls and have watched other people struggle with simple situations yet navigate complex challenges with more ease. Sometimes people figure it out and sometimes they crash into walls that were obvious to a savvy spectator. Through it all, I have honed my own way of observing situations, polished my own decision-making process, and learned how to help others not only see their situation in a new light, but also help them to make what they believe to be the right decision.

The end result is that my career has now led me to focus on what I love and what I believe I am the most effective at: helping companies through their first steps of their commercial life and helping them build teams that provide the initial foundations of what we hope will become a significant company. A few years ago, I started informally sharing my learnings around market entry concepts as a mentor to incubators, accelerators, and to the startups where I served as an advisor or a board member. One of my incubator founder friends challenged me to take this ad-hoc material that he found valuable and turn it into a more comprehensive course format focused specifically on business-to-business (B2B) market entry challenges. Through many iterations of giving the seminar, listening to the questions and feedback from participants, and thinking about illustrative examples, I developed a real class called the "Market Entry Strategy," and over time and many presentations, the materials became more polished and refined. The pandemic impacted my seminar significantly as the audience grew from a few tens of attendees in each class to an unlimited reach through the power of webinars and I received broader valuable feedback. And so, by now to my own surprise, the class has been attended by more than five thousand entrepreneurs and startup executives.

The seminar has received a fair amount of praise from startup CEOs, who often requested not just a copy of my presentation, but something that they could use to dig deeper into the methodology. Throughout the class, I recommend a number of specific exercises for teams to execute to find their positioning, to create executive alignment, and to provide founding teams with measurable objectives, and I share with them different kinds of playbooks. The tenets of the methodology advocated are meant to serve as a guidebook for them as they take their young companies from formation to becoming solid, expanding businesses with growth potential. This book is intended to serve specifically that purpose—to provide a reference that seminar participants wanted, and to provide other readers standalone material to help any startup team embarking upon their journey. The steps outlined are equally relevant for teams that are launching a product in a more mature company, and many of the seminar attendees fall into this category.

Many of the elements here represent well-established best practices and I do not claim each on its own is innovative. However, what I have attempted to do is to assemble previously scattered pieces of entrepreneurial wisdom along with my own insights into a rigorous and usable order that will help you anticipate and prepare for what you will face, and help you navigate around the pitfalls that have caused others before you to stumble. My best recommendation for how you should use this book is as a framework for setting your course, starting with your earliest team brainstorming sessions, and using it as a reference guide as your company continues through its formative years.

Note that this book is written to be read from beginning to end. Some books are written where the reader can skip around and pick topics that interest them, but the Market Entry Strategy methodology is presented in a way where each new topic is built on previous topics. Unexpectedly, quite a few entrepreneurs have attended the seminar several times, as each time they find another nugget that is relevant to the stage they are at, and you might find similarly that certain parts resonate more when you reach their associated steps. Many attendees have also reported making course corrections, refining their positioning, changing their sales approach, and especially renewing their focus on team alignment and execution as their company grows.

If you are a startup team early in your founding, or a commercial company trying to grow your revenues, or if you are a product manager at a global corporation about to launch a new product, then I have written *The Market Entry Strategy* with you in mind. I hope you enjoy the book and that it serves its purpose as a companion travel book on your entrepreneur journey.

PART I

FOUNDATIONS

1

A New Approach to Market Entry

"Wherever you see a successful business, someone once made a courageous decision."

—PETER DRUCKER, management consultant and author

After teaching my Market Entry Strategy seminar to thousands of entrepreneurs, one key feedback stood out: there are very few practical advice sources available for startup CEOs and management teams navigating this critical early phase. With a few notable exceptions, I also find that a lot of startup advice is being offered without firsthand experience and there is a huge gap between theoretical principles and the experience of standing in front of an all-hands company meeting on a Monday morning explaining the decisions you made over the weekend.

Many books on startups and entrepreneurship were written a while back, and although I have enormous respect and gratitude for their formative impact on the startup ecosystem, some of the principles of popular books have become outdated or are sometimes misused. As an example, *Crossing the Chasm*, the most popular of all books for company founders, was first written when startups were a fringe phenomenon looked at with suspicion by many organizations. The underlying concept was that the best market entry for startups would be to find and target early adopters whose influence would then carry a startup product into mainstream adoption—across a chasm where many young companies struggle. Investment capital has been flowing at two orders of magnitude higher intensity than in the 1990s, and startups now pile into the obvious markets in hordes. Large companies have gone from avoiding working with startups to fully embracing them because they understand that working with startups and seeking access to new technologies is critical to their competitiveness and sometimes survival. Examples of large companies whose business didn't survive, like Kodak, Blockbuster, Nokia, and many others, are etched in their collective corporate memories. The new landscape facing startups of today is not so much finding early adopters but standing out in an ever-growing crowd of look-alike companies. The biggest challenges today are having a visible, sustainable competitive advantage and a differentiated positioning that allows your product to stand out. It is about finding your way to market leadership before others, which often means seeking being crowned a leader even before the market has been fully defined.

Another popular approach that has been influential in entrepreneurship, and that also has a wide following among startups, is what's referred to as "The Lean Startup." The methodology embraces a model that has been promoted by accelerators and incubators in the entrepreneurial ecosystem where startups are encouraged to quickly develop a minimum viable product (MVP) and put it out in the market to get rapid feedback. These short cycles fit the process of the incubators and accelerators, where companies graduate from their programs on a "demo day." Then, based on an early market response, the startup constantly experiments, iterates, and refines its product. While this approach matches the fast pace of these environments, it fails when startups are pursuing transformative business-to-business offerings or deeper technologies because those products often require longer development time and deep market knowledge. While this approach has provably helped produce many "unicorns," it has also produced a large number of forgettable startups and myriads of burned-out entrepreneurs. Though inefficient, this approach is still at the foundation of many incubators and accelerators that achieve above-average returns for investors. But the social cost and mental health impact for the entrepreneurs that result from the large failure rates can be questioned.

The Market Entry Strategy can be seen as an alternative or additive approach to the information presented in these other business books. The methodology is based on the notion that most, if not all, decisions are entirely yours to make and proposes disciplined analyses, deliberate course decisions, and organizational rigor to deliver focused execution and team alignment as your fundamental North Star. It is an approach that is adapted for today's environment where *differentiation* and *market leadership* are your actual initial goals.

> When I present this content in workshops and seminars, the responses I get from the more experienced entrepreneurs attending are, "Your methodology definitely is correct and will positively impact outcomes for those who have the discipline to follow it," but also, "It is absolutely the right approach but you really need a lot of discipline and focus to follow this methodology." Both are true and I hope that this book serves as a guide for what to do—and what not to do—at your market entry stage.

The Market Entry Strategy methodology consists of a series of steps—each built upon the previous. Each step includes exercises you can use as foundations for your brainstorming and decision-making, and anecdotes in text boxes are used to illustrate a point under discussion. If you take each step seriously and do the recommended work and the exercises provided, I can only promise you a better understanding of your reality as a startup; a reality that involves facing head-on the challenges, battles, and surprises. While I don't expect that most people will abide by all of what is being proposed, I hope that the existence of a reference framework helps people understand how close or how far they are from a disciplined approach. At the very least, the Market Entry Strategy provides startups with a benchmark on their current progress against objectives and a set of practices to move their company forward.

KEY POINTS

- The landscape for startups has changed and the goals are now differentiation and market leadership in a crowded startup ecosystem.

- Market leadership can be achieved through a set of deliberate decisions and rigorous, aligned execution.

- The Market Entry Strategy provides you with a methodology and the framework to help you with the decisions you need to make to successfully transition to a scalable business.

2

The Market Entry Stage

"We are our choices."

—JEAN-PAUL SARTRE, philosopher and novelist

At a high level we can define the market entry stage as the moment when your first product is nearly ready to launch—or at least you think it is. Your startup is beyond the idea stage, the slideware stage, the user interface mockups, the concept demos, and the pilots. There is an actual product that exists that you can turn over to real users without your handholding and supervision. The market entry stage is also the point where a startup is now starting to talk to customers about using their product and about buying it. The startup is now at a point in their journey where they can start making headway in turning those users into paying customers and lay the proper foundations for future growth.

The market entry point is also where a lot of entrepreneurs get surprised because there is often a disconnect between the responses they heard from people when they were talking about their product and the sudden gap they discover between that praise and the difficulty in securing paying customers. Startups often receive what's called "happy air," which is the unbridled enthusiasm of potential customers for their product. Early-stage founders will talk to prospects, maybe conduct a broad survey, and receive quasi-unanimous responses where people love their product and encourage the startup to keep developing it. Venture investors often encounter this enthusiasm when startups pitch and they say, "I talked to three hundred customers, and they all said my product is amazing." Of course they will say it's amazing—why wouldn't they? But that does not translate into them actually being a fit for the use cases you will seek during your market entry, or issuing a purchase order, or a customer agreeing to dedicate a large amount of their team's time to actually trying to use the product in a real setting. There is an old saying among venture investors: "There is infinite demand for a non-existing product." Reading too much into what people say before your actual market entry can spell disaster for a startup.

> **Be cautious of endorsements from prospects or industry experts that are vague, such as saying you're addressing the right topic or praising your approach. Unless they are customers who have used your product, this is actually a sign of weakness or that you don't understand the bar you have to clear. Similarly, having many well-known investors or board members without relevant industry expertise can be a sign that you are seeking the wrong kind of validation, similar to the pitfalls seen in high-profile failures like Theranos.**

One reason for the disconnect is that a lot of startups get feedback in a non-traditional way, through personal connections or investor networks. People you are reaching through networks or connections are inherently biased because of the way you were introduced to them, and the feedback you receive from them is often misleading and won't help you figure out what it takes to find and acquire actual customers in the market at large; that learning process is key for the foundation of your company. This is also true of the concept of design partners that has emerged in certain verticals and some highly interconnected ecosystems. Design partners can accelerate your product definition if you choose them based on their fit to your technology benefits and a set of narrow use cases that map to your competitive advantage. If they are selected merely on your ability to access them, you could end up with a half dozen different use cases and different customer profiles. Even if a startup has a few paying customers acquired in this manner, if their market entry has not been done in a way that is truly representative of their ideal customers, there will be a delayed reckoning with the real obstacles to overcome and creating a repeatable motion will be difficult.

Although the decisions you make on market entry are critical, this is a point where many promising companies miscalculate. Startups often believe they can avoid making the actual decisions because they're such a hardworking and dedicated team or because they believe that ultimately the market feedback will decide for them, but the cost of that approach is often crippling. The opposite of experimenting in the market, of trying all sorts of things to see what works, is to be methodical and rigorous. Hence, if you are at the market entry point, you must actually figure out which market to enter, and you will need to first think about the following:

- How will you determine who your best prospects are, and how will you reach them?

- How will you position your product, and what messages do you need to communicate to which audience?

- What business benefits will you deliver and how will you know these are the ones that matter?

- How will you know that you are competitive and differentiated in your entry market?

- How will you price your product and what sales motion will you deploy?

- At what pace can you expect to grow, and how much capital will you need?

You need to ask and answer these questions and be able to build a clear plan to formulate *your* Market Entry Strategy. It is now up to you to make these decisions correctly and the choices you make will influence almost everything that will follow.

ENTRY MARKET DIFFERENCES

Every entrepreneur understands that the stakes are high in these very early days of your market entry. But despite the importance of this stage, not every startup does an equally good job of navigating it. Out of the several dozen companies I have invested in or worked with, I can think of three that illustrate well the critical importance of a successful market entry because they started out similarly and yet they each experienced dramatically contrasting outcomes. All three startups took about the same length of time to reach the market entry stage; each startup's CEO and executives were smart, dedicated, and hardworking. Each also had developed good products with unique benefits, and all three assembled quality teams. I had no reason to believe that all three wouldn't scale similarly and lead to robust growth and successful exits. But I was wrong about that.

The first company had to raise only one funding round and from there it quickly grew revenues and became profitable. When it was nearing $100 million in revenue, it was acquired for a very large sum and that provided an amazing return for the team and the investors. The second company launched its product to a more moderate reception and had slower revenue growth. It had to raise five financing rounds and reached $30 million in revenue when it was acquired for about half the amount the first company received. The product continues to be used today and while the returns were solid, they were much less spectacular compared to the first company for investors and the team. The third company, unlike the first two companies, almost always had to struggle. It had to work hard to raise a second round, having struggled to find a first solid foothold, and consequently had even more difficulty with a third round that had to be led by insiders. When it was eventually acquired, it received slightly less than the money invested in it and rewarded employees and founders with only symbolic sums.

As noted, all three companies were similar in opportunities, talent, and work ethic, but what made three apparently similar companies achieve such dramatically different results? The answer, after years of hindsight and experience, is now clear to me: the first company selected a narrow entry market in a vertical where it enjoyed unique competitive advantages, quickly established market leadership, was able to sell with high efficiency, and grew profitably shortly thereafter. The second company also identified a good entry market, but the business benefits of the company's product were less clearly defined and more difficult to communicate to customers. For this second company, every step of the sales funnel progression was less effective; each required significant expenses, and only modest sales were achieved. The second company also grew, but its less-efficient model produced a slower rate of revenue ramp and it experienced some churn, both contributing to a higher burn rate.

The third entered a wide and highly competitive market where it failed to establish leadership and was unable to differentiate its product from other offerings with similar messages. Even though the entry market it chose was rapidly growing, the third company wound up with little pricing leverage and struggled to emerge from the pack. It had to work extra hard to even be included in competitive bids and seldom won them. It constantly lived from hand-to-mouth, creating a deadly spiral where its product struggled to keep pace with its competitors.

These three companies were similar in many ways, yet they experienced dramatically different results. The critical way they were fundamentally different from each other was not in their product or team, but in the market entry strategies each chose.

IMPACT ON CAPITAL EFFICIENCY

As just illustrated, once your startup passes through the market entry point, your strategy and execution will determine your growth trajectory. That trajectory will in turn have a predominant influence on your *capital efficiency*. A successful entry such as that executed by the first company will naturally favorably impact your capital efficiency because you will experience significant growth for the investment you received, and if needed, you'll have an easier time raising capital at a high valuation. You are likely to need less capital as well, as the first company in the above example illustrates, because your initial success in your entry market will make you more efficient in executing your follow-on steps. It will set you on a favorable trajectory that will influence the foundation of your growth.

A side-effect of such explosive growth is that key industry players will take notice, opening the option for early mergers and acquisitions (M&As). Large companies know how to take such products and push them through their established channels where they can leverage customer relationships and accelerate sales without needing further proof of market adoption. During my career, I have worked with several companies that were acquired for extremely high multiples right after their initial market entry. They landed lighthouse customers, signaling that their innovative product would usher in next-generation technology, and that immediately put them on acquirers' radar screens. Though not necessarily the highest outcome, the relative capital efficiency from inception to acquisition leads to high multiples in a short time.

Outcomes for Different Market Entry Scenarios

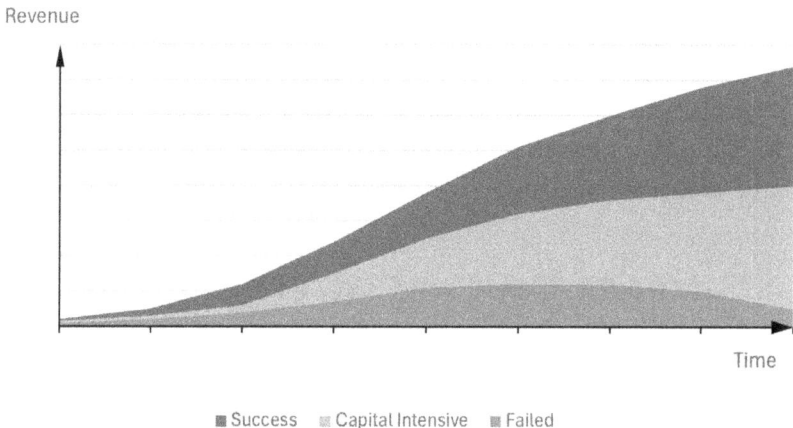

Revenue

Time

■ Success   ■ Capital Intensive   ■ Failed

On the other hand, a tepid market entry might still result in successful revenue growth, but at a slower pace. Like the second company mentioned above, you are growing with suboptimal capital efficiency. This requires additional rounds of capital and because of the sluggish growth, the subsequent valuations increase slowly, and cause additional dilution. The barriers you already have to clear are more costly, and there are now more of them and thus, your life becomes more difficult. Your company's lack of capital efficiency will be capping off the returns for your eventual outcomes.

But it could be worse. If your initial market entry fails altogether like the third company, you will likely not have met revenue and performance goals to justify an attractive round of financing. The post-money valuation of your last round may be higher than a new investor will be willing to accept in a follow-on round, resulting in a highly dilutive fundraise, assuming there is even any remaining appetite for further investment. Companies in these situations rarely survive. And when they do, they continue to be encumbered by the baggage from those initial missteps. They face dilution, demotivated early employees, and early-stage investors who are reluctant to participate further. They are likely to accept any outcome, even below invested capital, but in many cases, these companies will eventually die. Some extraordinary entrepreneurs have successfully recovered from such disastrous starts to achieve meteoric success in the same entry market or through a complete pivot. While such turnarounds generate inspirational media coverage, they occur rarely. It's far more common for companies that fail in their market entry to simply disappear without a whimper.

KEY POINTS

- The market entry stage starts when your product is ready and you can give it to customers without handholding.

- A successful focused market entry execution impacts the trajectory of a startup for the rest of its life.

- Companies that excel in their entry market receive numerous advantages that translate into superior capital efficiency.

3

Taking the Beach

*"You are about to embark upon the
great crusade toward which we have striven these many
months, the eyes of the world are upon you."*

—GENERAL DWIGHT D. EISENHOWER,
Supreme Commander of the Allied Expeditionary Force

Saving Private Ryan, one of my favorites movies, is mostly remembered for its unforgettable and gruesome opening scene showing the WWII invasion of Normandy. There is an entire fleet of carriers and battleships approaching the shores of northern France and then we watch as thousands of soldiers disembark to attack the beaches. Entrenched in fortified bunkers, atop difficult-to-climb dunes, we watch enemy machine guns begin ripping into Allied soldiers as they approach the beaches in landing crafts. More soldiers are shot in the water before they even landed, and still more are mowed down as they emerge onto the beaches.

The Normandy invasion remains among history's most expensive military engagements and one of the most critical victories of all time. If the Allies had not taken those beaches, history would not remember it as the entry point for the liberation of Europe. There were other, less-successful operations which the world has forgotten, like the failed Dieppe Raid of 1942 or the long and costly siege of Anzio. On the surface, it may seem to you that the taking of a beachhead in a war has nothing to do with your startup or its market entry, but I see many similarities from which there

D-Day Landing

are critical lessons to be learned. I always refer to the *Saving Private Ryan* beach-storming clip at the beginning of my presentations of the Market Entry Strategy to illustrate how daunting that stage is and to impress upon the audience the difficulty and the importance of executing it well.

It is the single best example I know to draw an analogy—and of course in no way a comparison with the actual sacrifices—to what a brilliantly executed market entry looks like, how painful a market entry is, and how singular focus on achieving market entry is key. The market entry step for a startup is tantamount to conquering the first beach on D-Day. It is the point from which you can establish your first base camp and where you can establish the makeshift harbors that will be used to bring in reinforcements and launch the subsequent attacks. And if you are not efficient and focused, you will linger at this stage with the equivalent of high casualties—in your case, a high burn rate relative to your progress. If you fail, you will be thrown back into the ocean.

The key element to remember is that the Allies didn't attack simultaneously the beaches of Brittany and the shores near Bordeaux. They focused their entire gigantic arsenal on one single point so they could bring to bear all their firepower in a single location to maximize the likelihood of having a first solid landing point and not get repelled to the seas. As a startup, you are going to have to pick a beach; you have to concentrate all effort on a single point and successfully conquer it. We will see how relevant the analogy will be throughout the book.

KEY POINTS

- D-Day landing is a great analogy for the market entry phase of a startup.

- As a startup, your first priority is to identify your "Normandy Beach," the entry point in your market.

- You will need to gather and focus all your resources and firepower to succeed in this market.

- Startups can use the imagery to picture the difficulty facing them and the need for an uncompromising resolve and discipline in their market entry to be successful.

4

Why Is Market Entry So Hard?

"Some things must be believed to be seen."

—GUY KAWASAKI, entrepreneur, evangelist, and investor

Though it might seem shocking to use the *Saving Private Ryan* scene and the associated analogies, it serves to reset expectations of entrepreneurs and prepare them for the tough battles they might not foresee. In this chapter, we will review some of the key challenges.

UNWARRANTED CONFIDENCE

Why is there such a massive difference between early assurances given in meetings and wallets being opened to purchase products, not to mention actual deployments and end users singing your praises or serving as favorable case studies? Nearly all qualitative data gathered by startups contains false positives. It's not that your prospects and industry gurus intentionally deceive you so much as old-fashioned human nature. Most people who gravitate around the startup sector like to encourage entrepreneurs, especially when they show you their heart-and-vision-filled presentations. It's easier to be supportive than to deflate their expectations and easier to agree with people rather than enter into debates about the value of the benefits or the uniqueness of their solution.

False positives crop up from other people, too. Often, professional industry influencers and luminaries have different scripts than your prospects, and they too are usually sincere in their encouragement. But influencers and luminaries make their living by spotting early trends and their own status and marketing depend on being quoted and having their ideas and posts shared by others. The media is always hungry to tell readers and viewers about early trends but again, that doesn't necessarily lead to purchases of your product.

Even real prospective customers can also unintentionally mislead a startup. Potential customers are typically skeptical of purchasing any new product because they're focused on trade-offs between benefits and costs—and any product you show them will obviously impact that calculation. But, similar to influencers, potential customers are also happy to meet with startups and are prone to tell them how much they like what they are up to. Customers may even be willing to try out one of your products for evaluation. But despite all those promising signals and sincere praise, prospective initial customers will still be reticent to be among the first to buy and deploy your offering. Another variation is that those friendly early customers come up with requirements that are not aligned with your actual business benefits, but since they're engaged and you want to satisfy them, you'll likely let your product roadmap drift away from its core focus to whatever they are asking. You will mistakenly think of this as progress but it is quite the opposite.

The last group of people who often provide false positives are angel investors. Here again startup founders are quick to name-drop famous entrepreneurs as investors in the seed stage but almost without exception, many of these famous angel investors made a snap decision over a lunch, a cup of coffee, or a quick conversation at an event. Of course, most of these successful entrepreneurs tend to have great investment instincts and many of them have relevant market knowledge but for them, deciding to invest $100,000 behind a startup's project is hardly the result of many hours of research and diligence. For many of them, it's really a small part of their assets that they reserve for "play money" and for the pleasure of staying involved with leading-edge technologies. Savvy investors know how to recognize those endorsements for what they are. It is acceptable for startups to take comfort in those high-profile accolades but to also be aware that they might not help in your entry market. What you really want are a few quality endorsements from people who understand your market and have spent the time to dive into your product at a deep level, not just famous people.

In all of the various forms, false positives can lead startups to overestimate their position. While endorsements, praise, and promises of deployments certainly elevate morale and can generate useful word-of-mouth marketing and initial awareness, those types of feedback are misleading at best.

COMPETING AGAINST CUSTOMER HABITS

Let us look deeper at another reason why customers are reticent to go from the pre-product endorsement to making an actual commitment once you are at market entry. Over time, I have found that the top reason is simply that old work habits are hard to break. While most entrepreneurs would like to believe they solve a critical problem—the proverbial painkiller—the fact is that for the most part, customers were doing just fine before your startup came along with a product that will intrude on their regularly scheduled lives. That doesn't mean your product won't increase your customer's revenue, reduce their cost, or enable them to tackle missions that were previously impossible. It just means that they were doing fine before you came along.

Perhaps that customer is getting by with work-arounds, with Band-Aids, or by throwing human resources at a problem that should be automated. Maybe your potential customer is avoiding a market that could be valuable to them or there is yet some other factor you're unaware of. What's important to understand is that this company is an established and going concern, operating every day and doing just fine without you. So, you will face some resistance when they look at how much change they must make to enjoy the benefits they told you they were keen to have. And the hard truth is that nearly all customers are naturally inclined to continue doing what they have always done.

I was on the board of a company that had a breakthrough
product that would enable a large corporation to whisk
away a $7 billion market from their main rival. We had an
independent third-party lab testing that validated that
fact and nobody questioned the validity of our claim.
I had a long-term relationship with their CEO and obtained a meeting with the top executives to present the
opportunity. Everyone was effusive about the joint solution and the CEO committed on the spot to move forward. Several weeks later, we heard from a director-level
person that in six months they would come up with a list
of possible customers for the new product to gauge the
market receptivity. Which of course never happened.

Do not underestimate how difficult it is for a startup to pierce through
established work habits. I have seen startups literally provide the solution
for businesses that are heading into a wall and about to collapse without
the new product being offered, but because the decision-makers had entrenched habits and aversions to disrupting their floundering businesses,
they were still reluctant to embrace a new offering. Academic research
corroborates how reluctant people and organizations are to change. You
may already be familiar with Clayton Christensen's classic book *The In-
novator's Dilemma*, in which he explains how companies remain generally
reluctant to attempt even the most minimal actions that could disrupt
an existing revenue stream for a hypothetical dramatic gain down the
road. So, one more reason that market entry is so difficult—beyond false
positives—is the very real issue of entrenched habits.

INVISIBLE COMPETITORS

The third most common obstacle startups face comes from well-established competitors who may not even show up in any market or competitive analysis that startups conduct, or any of the four quadrants on

your market slide. These are the existing suppliers to your prospective customers in adjacent markets. These *invisible competitors* can take you by surprise, raising your cost and time-to-sale. Even if you have an incredibly superior product, the current suppliers to potential customers have the formidable advantage of enjoying long-standing relationships with your prospects.

Who knows if you are going to actually be around three years from now? And even if you are around for a while, will you even remain an independent company? Why would any sane decision-maker risk their career on your dream? Why should an executive personally push and drag you through their maze of bureaucracy, purchase request justifications, return on investment (ROI) calculations, security clearances, and budget reallocations for the promise of a yet-to-be-proven innovation when perhaps it can be supplied by a trusted vendor?

While you busily exhaust yourself trying to get your foot in the door, all your incumbent competitor has to do to stall you is make a presentation that recognizes the market trend you've been highlighting and the technology direction you've been pitching. They can even compliment your product originality and early market traction. Then, of course, they will reveal a roadmap showing that they are working on solving this problem with promises that they'll have a good-enough competing product eighteen months down the road. Why should the customer suffer the risk of bringing a new vendor onboard? And by the time you would have effectively ramped up, the potential client might be only a few months away from being able to receive the same benefits from their trusted vendor. It seems like an unattractive proposition for anyone but the most maverick of corporate executives—and those people are rare within established companies.

UNPROVEN BENEFITS

Finally, the most obvious but often overlooked barrier you face as you struggle to gain traction is that *you need to have customers to get customers.* As you knock on your first door of a potential customer, you do not have a single customer who has already deployed your product and other than your word and sometimes your prior reputation, no one can attest that they can rely on your product viability and point to its demonstrated benefits. Market entry is the classic chicken-and-egg scenario, and the startup must somehow find a prospect who *believes without proof* in a miracle they cannot touch, evaluate, or measure. In short, you are asking them to believe it before they see it; to trust you before you've proven yourself trustworthy.

In today's fast-moving environment, an executive's career or even a technical professional's next opportunity can be made or destroyed because of a single, highly visible decision. A security breach that results in millions of stolen records, for example, or a production environment that collapses and knocks entire businesses out of commission, or associating with a business that damages the environment or is found to use discriminatory practices are all highly visible. Any of these can lead to enormous losses, tarnished reputations, and have career-ending consequences. And any of these is enough for previously high-flying decision-makers navigating obstacles to abruptly find themselves in the hot seat.

Of course, a brilliant technology adoption or strategic business transformation has a dual and almost equivalent benefit to propel an executive's career forward, but people will almost always prioritize self-preservation over a possible unproven gain. Many products need to be installed by the customers before they can prove themselves. With numerous technologies based on statistical models or machine learning, there's an actual lag time between the installation and the generation of relevant results. In other cases, when products promise to deliver efficiency gains, a customer needs to install the product, use it for months, and then make a hypothetical ROI comparison to what would have happened with the status quo (if they can do a proper side-by-side comparison).

The underlying and often hidden dangers for startups are pernicious—and preparing to overcome them is one of the critical points of the Market Entry Strategy. As a startup, you need to adopt a unique position to that first battle, which has been suggested as somewhat analogous to D-Day. You need to configure yourself so that all your forces are pointed at the one beach you decide you will land on. The startup equivalent to that specific order of battle is called *spotlighting*.

EXERCISE

MARKET ENTRY CHALLENGES

1. List the different forms of validation you have received and highlight how they can make you overconfident.

2. Describe the current customer habits that you will have to overcome.

3. Enumerate all the well-established invisible competitors.

4. Write down expected benefits you will have trouble demonstrating until you already have customers.

KEY POINTS

- Market entry has inherent obstacles that are difficult for many startups to anticipate and that also represent unforeseen barriers.

- The most common challenges of market entry are false positive feedback, entrenched customer habits, invisible competitors, and unproven benefits.

- Startups that identify these common challenges before their market entry can configure themselves to be in a position to overcome them and be better positioned for their beach landing.

5

Spotlighting

"In order to be irreplaceable,
one must always be different."

—COCO CHANEL,
French fashion designer and founder of the Chanel brand

Startups face a number of seemingly intractable obstacles; if you are starting a company, you cannot help but be discouraged by these. The odds really are against you. The classic and typical entrepreneurial response is to work harder. There are countless stories about entrepreneurs who knocked on one hundred doors and had ninety-nine of them slammed in their faces until they finally got that positive response that changed everything. Stories like this are embedded in the Silicon Valley zeitgeist and propagated in all the world's innovation centers, accelerators, MBA programs, and by the business media. Knocking on one hundred doors looking for that one customer who finally opens up to you is just like random assaults on the wrong beaches—exhausting, discouraging for the founders and team, and capital inefficient. Although some amazing legendary companies have been built that way, for countless others (and especially significant B2B companies), this iterative approach produces random and suboptimal market entry outcomes.

In contrast to those approaches, the Market Entry Strategy not only recognizes that the earliest stages of a company are unique, but that the circumstances startups face at that time call for a specific approach to develop a winning strategy. A brief glance at world history shows the problems experienced by military operations where an army will not concentrate all of their forces on a single point, and when that happens, they seldom make much progress. That is in fact the equivalent of knocking on every door or iterating countless times to see what the market chooses. Circling back to the D-Day analogy, this would be akin to the Allies waging a multitude of attacks on the 1,000-mile western coast of France with the hope that one of them works. Instead, they configured themselves in a unique and focused way for that stage of the battle.

The Market Entry Strategy uses that same idea for a startup market entry and the analogy used is called "spotlighting," borrowed from the language of theater and live performance. In fact, think of the opening moments of any theater performance. Before it starts, everyone in the audience is engaged in their own conversations but when the lights dim, the room becomes silent and then a single spotlight turns on. There is sustained silence as all eyes in the theater turn to where the spotlight focuses. Everybody in the audience will look at that person in the beam of the spotlight—everybody looks at the same thing and by controlling what the spotlight shines on, the director sets the initial focus of the performance.

But what would happen if some avant-garde director decided to start differently, and instead of using just one spotlight, they had every spotlight turn on simultaneously? Under each light, solo performers appear in unsynchronized individual routines. The audience would become immediately confused—they don't know where to look. And, depending on where they look, different people in the audience will see a different performance.

At the end of the performance, people will tell each other a different story because they will have seen something different. The resulting word-of-mouth propagated will be very confusing because their descriptions and impressions of the performance will not match. Conversely, if everyone walks away with the same story and talks to others, that story will resonate and amplify the word-of-mouth effect. Illuminating the stage with multiple spotlights is obviously uncommon at theater and live performances, but all too common among startups—and is exactly the wrong thing to do. It is the opposite of what happened in Normandy, and it is the opposite of what the market entry stage requires. Positioning for market entry using the spotlighting analogy consists in laying out a scenario where, as you enter the market, everybody is looking at the same thing and telling the same story.

Similar to the director of a live performance, as a startup team *you need to decide what you put under the spotlight* because you only get one spotlight to work with. You must choose your one story and make sure you communicate it to the people you think matter the most. Your spotlight shines on just one thing so people know where to look. It is far more effective than letting the audience choose which of the different parts of the performance to watch. Just like in live performances that begin with a single spotlight focused on one character, place, or activity, people will talk to each other about their perception of your startup and if they tell and retell the same story, this will create resonance and amplify your message.

> Frequently I get the following question: "If my company has general technology applicable to a couple of different verticals, do you see it as okay to have two or three spotlights, assuming that the underlying technology and benefits do not change?" My answer is always the same: "No, it's not okay. You can experiment, you can play with more than one market entry strategy, but once you have picked your entry market, you can't have two or three spotlights, you have to pick one."

When people create a startup, they usually are attracted by the total market potential, but focusing on the total market is different than what you need to do during the market entry stage. The Market Entry Strategy requires that you spotlight, which means narrowing your focus to your first market. Your market entry is your first curtain raise and you need to tell one single compelling story so that all the stakeholders you will address—not only your customers and market influencers, but also your management team, employees, investors, and your board—will all know what you have in the spotlight.

The idea of shining a spotlight on one thing and one thing only is a tough pill for many startups to swallow, but a startup must pick one because they really can't do more than one effectively. No matter how many benefits you have, no matter how many opportunities you have, you must get all your customers to focus on one thing and one thing only, on a single spotlight. And the key criterion you will use to decide what your startup needs to put in the spotlight is *differentiation*.

KEY POINTS

- To overcome the market entry obstacles, you need to put all your focus on a single spotlight.

- You get to decide what to put under the spotlight; you must choose the one story you communicate so everyone hears the same message.

- Regardless of all the benefits your product offers, having one single spotlight is your best strategy.

6

Differentiation

*"In a crowded marketplace, fitting in is a failure.
In a busy marketplace,
not standing out is the same as being invisible."*

—SETH GODIN, business executive and author

For a modern startup aiming to surpass its competitors, the spotlight must be the benefit differentiation, not how innovative it is, nor how novel the solution is, nor how the product is a game changer shaping the future. It has to only be differentiation. And the main reason for that is due to the evolution of the market. As technology innovation was coming to the forefront four decades ago, the school of thought on startups and entrepreneurship was focused on getting early adopters for a startup's product. Today, finding large companies to be early adopters is no longer an issue; the problem startups face today is standing out in a landscape crowded with similar startups. Which startup should a big company do business with? Large customers want to work with startups, they just don't know which one. They are being bombarded with all combinations of buzzwords by startups using all forms of marketing technology, using social media platforms and all kinds of automation to constantly try to get the attention of new prospects, and none of them stands out.

Today's market requires a different mindset, very different than sending tons of emails or writing countless blogs to see what sticks or generates traction. Instead, startups need to think carefully about their entry market and about their differentiation—that element that will bring attention to their company and product, but not to others. This is what you want to put under the spotlight. Note that this decision is *entirely in your hands* and hence you shouldn't abdicate that choice in favor of a limited, random, or biased experiment. You want to find an audience for your product, and you want your entire audience to focus on one thing and one thing only to maximize your success. Hence you need to find a way to *both select your audience* and *what you want them to look at*, which is the equivalent of choosing what to spotlight and whom to spotlight it for. Let us see which critical elements are part of that decision.

ENTRY MARKET VS. TOTAL ADDRESSABLE MARKET

As previously discussed, a common mistake that entrepreneurs make is often the result of their fundraising efforts. In many meetings with investors, they'll be asked for the size of the market and be expected to portray an extremely large opportunity. This is a logical question from investors since the size of the exit—whether it is an initial public offering (IPO) or an acquisition—is strongly related to the market that is opened by the company they invested in. However, even though the entrepreneur will also be attracted to the size of the *eventual market*, their first job in the early stage of a company is to land on a beach. Without conquering the first beach, as we saw in the historical case of Normandy, there is no conquest of France and no liberation of Europe. The huge market beyond the beaches of Normandy is Europe but for the Allies, their *entry market* was that beach in Normandy. Obviously, without winning their entry market, there would be no opportunity to lead the overall market.

So, it does not matter how large the overall market is unless you can also convince your investors that you have a method to penetrate it. Sophisticated investors will be more impressed—and more likely to invest—if you actually have a market entry strategy. For example, suppose your startup has an innovation that could benefit billions of people.

That's an opportunity that many investors could get excited about, but also a dauntingly large target. How are you going to penetrate that market? How are you going to convince billions of people to use your product? More importantly, whom do you start with?

Imagine if you had to decide on an entry market for a network-based communication system (for example, Skype, a precursor to Zoom)—where do you start? Recall that back then, there were no unlimited calling plans, international calls were expensive, and telecommunication companies would charge for each text message. Skype had to overcome the classic network friction: you need existing users to attract more users. The company analyzed the situation and decided to seed their user base with foreign students. They didn't just try to see who would sign up across the entire globe, they picked those foreign students as the entry market because they had the most pain trying to connect with friends and family and were more likely to sign up and bring others on. Having already a devoted core user group, it was much easier to expand the network.

One of the key elements in deciding about your entry market is that you will need it to be the one where you are sure to be *differentiated*, or simply stated: you need to be *unique*. That means starting with the entire market and cutting it into smaller submarkets and then hunting for the sliver where you, and only you, can address a customer need. Every startup will most likely have some overlap with competitors, but you need to find that use case where you can do something that others can't for that entry market. Maybe you address a certain size company, or a vertical, or a region. It doesn't matter what it is as long as it's the one market for which you deliver a business benefit that others can't. Once you shrink the market entry point to the one where you are unique, then you can put that in the spotlight.

The idea of whittling down the entry market is a very difficult thing for a lot of entrepreneurs to accept. It feels like you are cutting off alternatives that maybe, just maybe, could have been the paths to success. Shrinking the aperture seems tantamount to reducing the opportunity set. That is in fact a fallacy because you cannot pursue all the opportunities with anything close to the necessary vigor to make a dent. And so rather than reducing the opportunities, you are in fact choosing the opportunities that are the right ones. This is a terrifying choice—but a necessary choice!

How do you decide which market to choose? The Market Entry Strategy takes you through that decision process in the next steps.

EXERCISE

ENTRY MARKET VS. MARKET

1. Describe your Total Addressable Market (TAM). Make sure to use public, top-down information (this market is $14 billion and 12% Compound Annual Growth Rate (CAGR)) as well as a bottom-up validation (each customer will pay $20,000 and there are 4,000 of those).

2. What would be all the subsegments of this market?

3. Which of these would be relevant to you? In which do you have the strongest differentiation?

4. What would be the Serviceable Addressable Market (SAM) of that submarket, i.e. the part of that market segment you could address given some limitations such as geography, customer size, etc.?

5. How do you tell the story of your entry market vs. the total market?

KNOW THYSELF

The next element of the entry market decision starts with a *deep awareness* of your company's assets, who you are as a team, what your key breakthrough is, and what benefits it will bring to the world. By deep awareness I mean a completely candid, forthright conversation with yourself and your team. What are the true advantages of your product and what problem does it solve really well? This is the point where you need to be laser focused and resist the temptation to list everything that it might do some day or every problem for which it can bring an incremental improvement. Having this conversation implies a deep knowledge of the state of the art in the space you are attacking. I've seen entrepreneurs time and again throw themselves into the battle before having done their full homework and have often seen "demo day" presentations for which a simple Google search reveals large, existing competitors already serving quite well the same market. From my perspective, the best research process from startups is close to what I experienced years ago when I worked on my PhD. Doctoral candidates must develop an original thesis that advances the state of the art in their chosen field and then propose their thesis to a vetting committee before being allowed to conduct their research. Candidates have to put in a good deal of time—usually years—before they have enough proficiency in their field to be able to truly claim that their proposed research is a novel advance.

Similarly, a startup should have the most advanced knowledge of what is going on in its field, and will need to know existing products, what prior attempts other companies made, and what projects failed. If other startups or established organizations failed, it's easy to conclude that they weren't smart enough or that you will out-execute them, but that is naïve. Instead, a startup needs to be brutally honest and explain to itself (and others) why it will avoid the same pitfalls and avoid failing like other teams did before.

In doing this analysis, it is equally important to have a deep understanding of the team's DNA—whether they're world experts at graph theory, or have been pioneers of unsupervised deep learning, or know how to compress bits across a network, or maybe have decades of knowledge of automating mid-market customer acquisition or know the intricacies of the real estate market of suburban, mid-size cities. Having a solid grasp of the skills of the team will strongly influence the choice of the entry market. Likely, the DNA of the team will have an impact over the life of the company and influence for years to come where key innovations will come from.

This step (like others in this book) requires that the startup team lock themselves in a room full of whiteboards and notepads and, through an incremental refinement process, write the shortest possible list of their unique technology breakthroughs, the differentiation of the product, the DNA of the team—all of it. This is *not* a casual conversation over a beer or a forty-five-minute brainstorming session after a weekly meeting. Ideally, the list of fundamental breakthroughs should be whittled down to one, but if you have three identified breakthroughs, that's probably okay. Anything more than three is going to get you in trouble and you are likely to be tempted to shine multiple spotlights on multiple things. Entrepreneurs have told me that this part of the Market Entry Strategy methodology is excruciating, difficult, and emotional. But the startups that actually do a thorough job of conducting this important self-reflection are light years ahead of their competition. Two equally important positive outcomes result from doing this work, according to many startups: First, you will know yourselves and your team better than ever. Second, everyone who participates in this exercise will be fundamentally aligned. If you don't do it for the benefit of the methodology, do it for those reasons!

EXERCISE

TEAM DNA

1. What is the team background in terms of the following?

   - Areas of deep technical or market knowledge

   - Formal or informal education that is relevant

   - Professional experience providing unique insights

2. Recognize strengths/weaknesses and anchors of current/future innovation.

3. Highlight must-close gaps through hiring and the associated timeline.

4. Highlight the consequence of team DNA in determining your possible entry markets.

KNOW YOUR CUSTOMER INTIMATELY

The third element of this quest for your spotlighting strategy is to know your customers. I use the word "intimately" intentionally to signal that a startup cannot afford to just know them superficially. You need to acquire the deepest knowledge of their interests, their problems, their behavior, their timeline, their established habits, their current vendors (including those adjacent to your product), their security and regulatory concerns, and the list goes on. Just like you needed to do PhD-caliber groundwork on yourself, you need the same rigorous approach to understanding your customers. To know your customer intimately will take significant time and effort; the "Customer Intimacy Test" exercise below will be a good starting point. The purpose of this is not to add work for the startup, but because I have seen all too often very good startups fail because they didn't know their customer well enough. They didn't fail because they were not solving a critical problem, but because they became aware too late of an arcane but critical constraint of a customer. For example, if an industry must replace some components at a given interval for regulatory reasons, a startup with a new optimization to make components last longer might seem interesting but is ultimately irrelevant. If your new automated medical diagnostic can save patient lives but the workflow of the hospital has no ability to free up a doctor's time to look at your output, it will equally fail to be used. If you sell a product that needs to be designed into another product, you will only have the time between the customer product project launch and design freeze to sell—before or after that window, you are out of contention.

Ideally some members of the startup team have either been employed by one of those customers in your target entry market or have successfully sold products on a wide scale to those customers. If you don't have that type of intimacy, you should immediately hire someone who brings that customer knowledge into your startup team. This is not a consultant or an advisor but someone who is within the company and will bring their specific market experience to bear in every meeting. It won't be cheap, but it is infinitely cheaper to acquire customer intimacy by bringing on a new team member who's lived and breathed in the market than it will be for a startup to get that intimacy through trial and error. If you decide to take the trial-and-error approach, remember that your company is running a high burn rate and every month of missed key performance indicators (KPIs) moves you away from the ideal trajectory for the next round. I would not skimp on anything that can increase your customer intimacy.

You might think, "Well, I can read up on the industry, the innovations, and the space and probably get to the same level as someone coming from the industry." Maybe, but probably not within five years. Assuming that the 10,000 hours rule presented in Malcolm Gladwell's *Outliers* is correct, then if you spent forty hours a week trying to become an expert in an industry, it would take five years to become proficient. And that amount of time is suspect, since every industry is evolving and presumably you have other things that require your time for your startup. No amount of reading will be close enough to get you to expert level in your startup's lifespan.

During the first wave of cleantech innovation in the late 2000s, I became passionate about investing in the water-filtration market. For an entire year, I read every publication and analyst report, attended multiple conferences, and roamed the hallways of the relevant trade shows. I concluded that there was a specific investment opportunity—filtering the most polluted effluents—because it was a critical need and I identified a company with a demonstrated breakthrough. This investment failed for a reason that was well known to any long-term industry participant: CATNIP. This acronym means "Cheapest Available Technology Not Involving Prosecution." That is, in this industry, customers never bought anything unless they were forced to as the only alternative from being prosecuted and sued for millions of dollars for environmental damage.

What if you can't find anyone you can hire to bring that knowledge? This situation presents an interesting test of your startup because if you can't attract somebody from your entry market, that's a negative signal about the viability of your approach, product, or team. If no great talented person already experienced in and intimate with the field can be convinced to join you, you might want to consider going back to the drawing board.

EXERCISE

CUSTOMER INTIMACY TEST

1. On a scale of 1 to 10, how well do you know your customers?

2. How long have you been working with this set of customers?

3. Do you understand how your customers adopt technologies?

4. What motivates your customers to make decisions, and what business benefits matter most to them?

5. Are they able to find budget? How profitable are they? What is their growth?

6. Who on your team possesses deep customer knowledge and intimacy? Who has worked in this industry? Who has worked for providers to this industry?

7. Do you need to hire an additional executive to improve your customer knowledge?

8. Create two lists: A) What you know with certainty about your possible market entry customers and B) What you need to know about your market entry customers before you select them. What activities must you complete to close your intimacy gap?

KEY POINTS

- Your entry market is where you start and it is acceptable and expected to be much smaller than your Total Addressable Market (TAM).

- To pick your entry market, you need deep awareness of your company's assets, team strengths, your product breakthrough, and your unique business benefits.

- It is key to assess the team DNA for strengths and weaknesses that will either help you penetrate your entry market or need to be compensated for.

- To inform your entry market decision, it is vitally important to know your customers intimately—to know their interests, problems, behaviors, timelines, habits, regulations, current vendors, and concerns.

PART II

BUILDING YOUR MARKET ENTRY

7

Finding Your Entry Market

*"The person without a purpose is like
a ship without a rudder."*

—THOMAS CARLYLE, essayist, historian, and philosopher

Once you have the deep knowledge of yourselves and are satisfied that you have customer intimacy, you can now work on selecting your entry market. Notice again that your entry market is not the entire market, it is a subset of that market, and you will need to decide which subset to go after to maximize your chance of landing in it and, as we will see, to dominate it. When you select your entry market you are spotlighting one benefit to a small subset of customers that you have chosen and for whom that is unique. That decision is entirely yours to make—and there are many negative consequences to *not making that choice*.

Unlike seeking early adopters or iterating based on random biased feedback from the market at large, a startup's entry market is a deliberate choice. You get to—*you have to*—choose your entry market from among the many possible markets. You can control who, where, and when. You choose the entry market so that when you talk to a customer in this segment, you know that you will have the highest chance of being differentiated. You *must know* that the spotlight you have chosen will be the one that resonates most with those entry market customers' needs. And the most important criterion to choosing your entry market, customers, and spotlight is accomplished by picking the one that will make it the easiest to land on your beach. For your startup, that means the one where you will most easily overcome the market entry obstacles that make market entry challenging (see the previous discussions on customer habits, invisible competitors, and unproven benefits). How do you choose your entry market? It boils down to one simple and fundamental principle:

> *Your entry market is where you can most easily demonstrate your chosen benefits that uniquely address these customers' critical needs.*

The key elements of this one principle are critical:

- **Uniquely:** You must be differentiated and be the only one that can make the claim you are making.

- **Easily:** To better overcome the obstacle of needing customers before you have customers, proving your benefits must be as easy as possible.

- **Critical need:** Unless the business benefits are critical, there is not enough pressure to overcome the obstacles.

All of the painstaking work on both knowing yourself and developing customer intimacy is the lead-up to making the correct entry market decision. You will know you have chosen the correct entry market when you can honestly look at it and confirm that for the customers you chose, your product (and your product only) solves the particular critical pain point you have chosen. You need to have made a clear decision of which benefits to spotlight and have the intimacy to be certain that, given their environments and their constraints, these are the customers with whom it will be easiest to make an impact.

Deciding on a market entry point is the most critical and difficult exercise. If you discover other competitors that, in your entry market, can do what you can do then you are not *unique* in that sliver of the market.

If you find yourself in this situation, it's best to go back to square one and look at your benefits, analyze again your customer knowledge, and refine your thinking until you can absolutely state that you are unique. If you conclude that the problem you are addressing is not a *critical need*, you have to restart. And if you realize that this is not where it will be *easiest*, then look again. It is always preferable to invest your time in a startup tackling a smaller sliver of the market where it is unique and solves a critical customer need rather than invest in a startup pursuing a wider market approach where it will face plenty of undifferentiated, head-on competition. Similarly, you will have a hard time getting capital, as investors will avoid backing startups that will have insurmountable friction proving their benefits to customers or those that will simply end up at the bottom of their customer priority list.

I served on the board of a company that helped marketers easily obtain rights to images posted on social media. Initially, the company didn't choose a specific entry market, and while it showed early promise and rapid growth, things quickly slowed down, and churn became an issue. After analyzing the situation, we found that the travel sector—hotels and tourism—showed strong growth and retention, as they constantly needed fresh images. This insight could have been discovered from the start, and had the company focused on that entry market early on, it would have been more capital-efficient and likely achieved a far better outcome than the moderate acquisition it ended with.

I have been belaboring the importance of the spotlight and the singular focus of a startup on the one thing you can do that no one else can do, and at this point, you might be wondering, "Why is it so important to have an entry market?" The answer is that when a startup defines a small market that only it can address, it can quickly be seen as a leader.

Your goal as a startup is to have market leadership, and differentiation provides that leadership. If only you can do something in your entry market, then you are *de facto* the leader of that market. There is nobody else in the world that can do what you can do and for that specific sliver of the market, for that thing that only you can do, *you are the market leader*. Being the market leader means you start from a position of being number one. Now, you might be number one of something very small, but you are still number one. And the beauty of being number one, of being a market leader, is that you get a long list of valuable advantages, as will become clear in the following chapters.

EXERCISE

CHOOSING YOUR ENTRY MARKET

1. Enumerate the possible customer segments in your wider market. This is the equivalent of mapping all possible D-Day landing points.

2. Define the customer segment in the entry market you will go after. That is your Normandy beach.

3. What specifically characterizes the customers in this segment?

4. What makes you unique for them?

5. What unique business benefits are you going to deliver?

6. How do you know you are addressing a critical need?

7. What makes this entry market the one where you can most easily demonstrate your unique benefits?

KEY POINTS

- You get to choose your entry market and your initial customers. It is entirely your decision and most elements informing that decision can be known or researched ahead of time.

- You choose the entry market where you know you will have the highest chance of being differentiated, hence unique.

- You choose the entry market where you can most easily demonstrate your unique business benefit for these customers' critical need.

- It is always preferable to tackle a smaller sliver of the market where you are unique then pursue a wider market approach where you will face plenty of undifferentiated, head-on competition.

8

Positioning for Market Entry

"Leadership is the capacity to translate vision into reality."

—WARREN G. BENNIS,

scholar, organizational consultant, and author

The core thesis of this book is that the market entry point is the single biggest decision a startup needs to make. As was the case in the invasion of Normandy, successful startups will marshal all their resources and efforts on one sliver of a market where they can get a foothold. As a quick refresher, we have spent time exploring how to identify the right entry market, and how you will then highlight your unique differentiation in the spotlight—the one thing you want everyone to see about your company and its product. The spotlight will resonate with the small subset of customers who most benefit from working with you because you—and only you—are solving a specific problem that is a critical need for them. Once you have established your entry market, you will become the market leader because there is no one else that can do what you do. As a market leader, that position confers on you an array of valuable benefits which will become evident over the next few chapters.

THE FOUNDATION OF EVERYTHING

Now it's time to translate all this preliminary hard work into a series of talking points, documents, and visuals. These materials are critical for communicating both internally and externally, as consistently and concisely as possible, your differentiation. In this chapter, we will go through a formal process of building this positioning and develop the essential components of messaging. Positioning refers to where you want to be in the market, including where you stand vis-à-vis the key trends that are being discussed, how you are differentiated from your competitors, and how you positively impact your customers. Messaging refers to the precise language you choose to effectively communicate your positioning and clearly emerge as the market leader in the entry market you have chosen. Both are vitally important for the Market Entry Strategy to be successful.

An approach that works well for developing your positioning and messaging is based on the following sequence of steps:

From Vision to Plans

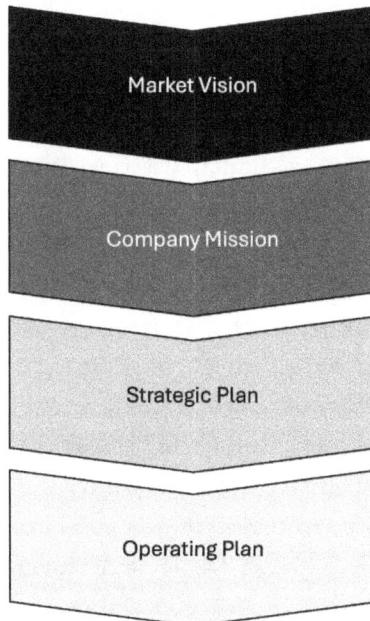

Market Vision

Company Mission

Strategic Plan

Operating Plan

Don't let the generic terms like "vision" and "mission" dissuade you from tackling this methodology because it's very important to dig deep on each part in order to fully own your positioning and your messaging. There is as much to be gained from the process as there is from the outcome.

This is not a founder-only exercise that you do in a room by yourself; instead, it's best if you involve your full executive team and get total buy-in for every step. You will likely discover that even among the core team members, there could be small or even significant disagreements. As you express ideas in writing (perhaps starting with whiteboards and notepads but eventually firming them up on official company boilerplate and other materials), differences between people regarding the exact problem you solve and the mission of the company will come to light. Conceptually, it's simple to understand this exercise but intellectually, it's a lot of work. The tendency is for people to say, "Yeah, I get it," but then to not do the work, to not write down the answers to the exercise. Yes, it takes effort, but you will be much farther along and in a better position than other competitors. This is the *foundation of everything*, and you cannot move forward and execute the rest of the Market Entry Strategy without a vision, mission, and strategic and operating plans.

Every word matters and every word needs to be exactly defined and consistently used. You need to try to relentlessly simplify every sentence, removing words that are not absolutely needed or replacing them with similar words that more accurately describe your precise positioning. Those exact words will appear on your website, in every press release, every boilerplate, every blog or post, every presentation, and every verbal communication. You should be striving for 100 percent consistency across all media and by all the representatives of the company. This consistency will give your startup a chance to have the market hear identical messages from people in your company, and everywhere else people come in contact with it. Having a consistent, concise message is the foundation of spotlighting, which is critical to maximizing your chances of being heard cohesively by all participants in your entry market.

Those words will be used for internal communications as well as externally and will be repeated over and over again in your weekly management meetings, your all-hands meetings, and your board meetings and will appear on every poster on your office walls. The complete buy-in of the narrative you create is critical because any variation of the words you have chosen will be reflected and amplified by others and will lead to a cacophony of messages that will confuse the market. If you are not aligned in talking about your startup using the exact same words, then you are undermining your spotlight and your approach is similar to all the spotlights coming on at the same time at the beginning of a performance: people will see, hear, and believe different things about the same company.

MARKET VISION

The idea of creating a market vision statement is to help convey to the entry market audience that you have an understanding and an opinion about what is going on in and around the market in which you intend to operate. It is a way to highlight your initial credibility with your audience, starting with customers, then investors, and finally with market analysts and influencers. In this part of the exercise, you start with a market vision that nobody is going to dispute. For example, nobody is going to dispute that the majority of compute workloads are moving from on-premise to remote centralized servers in the cloud. Or that cybersecurity attacks are increasing and that every financial institution or e-commerce company needs to spend a significant part of their budget defending themselves. Nobody will disagree that consumer time spent on mobile devices has significantly outpaced the time spent on desktops. The key point in this initial statement is that *the vision you state is non-controversial* and will elicit initial agreement from the audience you are addressing. Obviously, it needs to be relevant to your target market but general enough that anyone even remotely involved in the space will nod in agreement. By stating an obvious market vision that everyone will agree with, you are establishing a positive foundation to start getting people you don't know to be comfortable with you and the viewpoint you are about to formulate.

The next step in the vision statement is to highlight something that is happening in the space you just mentioned that is pertinent to what you are working on and represents an obstacle or a problem for the previously recognized trend. For example, a startup in the cloud infrastructure space could point out that security or data gravity (i.e., the difficulty and cost of moving large amounts of data) are impediments to the use of cloud-based workloads for certain applications. Or that in defending against cybersecurity, companies must still abide by local privacy regulations. Or you could state that despite decades of effort, there is no solution to the secure corporate use of mobile devices that doesn't affect user experience.

The goal with this more-refined statement is to set the stage for the audience to learn about the context of your startup—the sliver of the market in which you will operate and where your startup will improve the situation with the launch of your product. It is completely acceptable at this stage of the positioning unveiling to say something that's generic enough that you'll most likely not be original in stating the general trend nor the impediment—although you might be. The vision statement is not meant to highlight your differentiation, nor is it a forum for you to tell everyone how great your company is or how innovative your product is. It is only a statement that helps create a foundation for a common vision between you and the listener or you and the reader that will (hopefully) predispose them to recognize your thought leadership.

If you are at the stage where you're seeking funding and will be making a presentation to a potential investor, the vision statements—which are the general statements of the state of the industry and the context of your startup—should be your first slide. If you're writing content such as a blog post or a white paper, the vision statement should be your opening paragraph. If you're having lunch with a person who could be helpful to your startup, or you end up in the proverbial elevator ride with a key customer, your first words as you introduce yourself and your company should be the vision statement.

EXERCISE

CREATING A VISION

1. State the trends in the market that make your company relevant.

2. Describe an impediment the customers in your market face that your vision will address.

3. Validate that your vision is generally accepted. Are you getting nods when you state it?

4. Remember to refine these statements to the most concise and exact words possible because you will be using them for a long time.

YOUR MISSION

The next step in positioning your startup is to articulate your mission. The mission builds upon the previous market vision statement and the impediment, problem, or critical need you described. The mission simply states what your startup is going to do that will change the market outcome. You are not at the point of telling people *how* your startup will solve the problem, only *what*; your mission states only your startup's purpose and impact. What is going to be different in your market (and remember, we're talking about your entry market) if your customers adopt your product? What difference will it make? In the examples above, it could be that your startup will enable companies to securely run the most sensitive workload in the cloud. Or that you will effectively provide a solution to protect financial services and e-commerce customers from cyberattacks—at least of a certain kind—without needing to look at a customer's private information. Or that you finally have developed a way to allow secure corporate use of mobile devices without any impact on user experience.

This is the point where, having shared with your audience the framework of the market, you are now telling them something impactful that they are ready to accept or at least be intrigued by. They may have disbelief, they may be skeptical, they may dismiss what you are saying, but getting pushback can be a great sign because it predisposes them to hear the *how* that will soon follow. However, before you provide the how, it's best to underline the business benefits that will be gained from your startup executing the mission you just stated. By talking about the benefits, you will remind your audience of the issue you focused them on and what would happen if your startup succeeds—what would happen if you materially affected their outcome. For example, a company's workloads, now being secured, would run in the cloud and benefit from the lower cost structure, flexibility, and expandability that the cloud is known for. A company's financial services and e-commerce would enjoy fewer losses and greater trust from consumers. Or a company will finally be able to take advantage of the benefits of mobile devices without endangering corporate security.

Let me repeat, it is critical that your vision and mission statements are thoroughly infused in your team so that everyone is aligned, and everyone speaks the exact same language in the market. It's important to be ruthlessly concise and a perfectionist regarding the accuracy of the words because those words will be with you everywhere and for a very long time. Review your vision and mission statements again and again, challenging yourselves to remove words or replace them with fewer but more descriptive ones. It may seem like there are better uses of your time as a startup, that you should be refining your product or trying to make a dent in the market, or maybe hiring someone to write those statements, or even use AI. But there is no replacement for doing the thinking yourself—AI can't do it, an outside consultant can't do it, a Nobel Prize–winning writer can't do it. If you want your startup to succeed, the vision and mission statements must come from the concerted effort of the team. But once you do it, you will know yourselves, your team will be aligned, and you will understand your opportunity in deeper ways than you thought possible. Your vision and mission will help you on your way to becoming the leader in your entry market.

EXERCISE

CREATING A MISSION

1. State how you are actually going to impact the state of your market.

2. State how you deliver unique benefits that matter to your market.

3. Validate that your mission is unique, that nobody else could put their name above it. Try to insert the name of your competitors and confirm that they couldn't make that statement.

4. Validate that you are getting the audience's attention. If they are curious but doubt that you can do it, you are on the right track.

5. Review the words you have used in your mission statement, experimenting to see if you can replace a word with a more precise one. Can you delete a word without affecting the meaning of the statement or replace a group of words with a single, more accurate word?

KEY POINTS

- Positioning is key to successfully executing your Market Entry Strategy.

- Consistent and concise messaging and disciplined execution are the foundation of everything.

- Create your positioning materials so that you communicate your differentiation as consistently and concisely as possible.

- Your vision highlights your credibility and standing in the industry and points to indisputable facts and needs that you intend to address as a company.

- Your mission states what you will do to change the market outcome—the reason your market will be improved because you are solving a key problem.

9

The Key to Execution

"What we can control is our performance and our execution, and that's what we're going to focus on."

—BILL BELICHICK, football coach with six Super Bowl wins

Many startups immediately focus on their strategic plan, but the Market Entry Strategy advocates that spending time and effort to develop your vision and mission first will lead to better results. Once you have the vision and mission, you have a way to describe the overall market, know the entry market issue or problem you are solving, and are sharing how your startup will impact outcomes for customers in your entry market. The impact is now translated into what specific business benefits you would deliver. All of this work on mission and vision is setting the stage for the *what* your company and product do, and now is the time for you to articulate *how* you will execute your Market Entry Strategy.

THE STRATEGIC PLAN

The strategic plan details how your startup is going to get to the beach, how you are going to get a foothold in your entry market. When completed, the document should be shared widely with all stakeholders—target customers, analysts, investors, your own team, and employees. It should contain the goals you set such that if you succeed in your strategic plan, then you'll be able to accomplish your mission. It includes the strategic decisions in terms of focused entry market, key target customers, and how winning these will enable you to claim market leadership in your chosen segment.

For example, if your startup is serving the financial services industry, your entry market strategy could be to seek "mid-tier banks in the US Midwest region" because those are both easier to convince than the large banks and also more likely to be compelled by the solution offered by the cloud since they can't afford their own data centers. Even if you have a solution for "all mid-tier banks," your entry market is narrower: you choose a regional focus because you'll be more effective in concentrating your fledgling sales force on a well-defined target. Your strategic plan could call for turning into customers a number of those sufficient to start getting the resonating word-of-mouth conversations among a tight-knit community of regional bankers facing similar constraints.

In the case of protecting e-commerce companies from attacks while preserving privacy, you would perhaps decide to "focus on online pharmacies and life insurance companies" because both have access to highly sensitive medical information while at the same time being juicy targets for the cyberattackers. Or, in the case of securing mobile devices, you might identify "insurance brokers who are in the field" because they have the most compelling use case for using their own devices but still require a high level of security compliance. All three cases—financial services, e-commerce companies with sensitive data, and mobile devices that require high levels of security—have much bigger eventual markets, but you choose to focus.

Note that the most important part of the exercise is that it forces you to choose. It's your choice to determine how you'll select the customers with a critical need, your choice how to design your market entry experiment. The very essence of a strategic plan is that you actually have created one—you've spent significant time thinking it through and explaining to yourself why you chose one particular market over the thousands of others you could have chosen. If we go back to the Normandy invasion analogy, the Allies had to think, "Why these beaches? Why this timing? Why this specific combination of naval and air attacks combined with behind-the-lines para-commando drops?" They didn't think, "Oh, let's just pick a beach and see what happens." No, your Market Entry Strategy is a choice, and it follows from your ideas about the market, target customers, their critical needs, and how you can uniquely address them.

Many startups believe that their strategic plan is secret, that it shouldn't be shared with anyone beyond the leadership team, and that it should be kept highly confidential. They will often create all kinds of obstacles to prevent access to the strategic plan—even from investors—because they fear that competitors will get the plan and counter their strategy. Nothing could be further from the truth; as previously mentioned in the opening of this chapter, the strategic plan should be widely shared. As a startup, you want the freedom to tell the market openly and specifically whom you are targeting because by targeting specific customers, you'll find an easier market entry path. The mid-tier Midwestern banks, for example, will respond better if your message is tailored for and aimed specifically at them; they will recognize their problems, recognize their way of talking about them, and will see that you're intending to help them first. Conversely, the financial institutions outside of the market delineated in your strategic plan will not be targeted, saving you precious resources but also avoiding the unnecessary broadening of your entry market. If a global bank like Bank of America were to learn about your benefits and reach out to you, it would be a tempting but distracting waste of your time—so just tell the market what you intend to do.

If your competitive advantage is sufficiently slim that having knowledge of your strategic plan is enough for another startup to effectively compete with you in your entry market, then you were not differentiated enough. Since you chose this entry market specifically because *only you* could offer the specific benefits for the target entry market's critical problems, no one should be able to use your strategic plan—it is unique to you, based on your strengths, your team, and your vision. What about very large companies with their vast resources and market presence—couldn't they take a startup's strategic plan and act on it? Of course they could, but I can tell you from my own experience, that won't happen. You could meet with the CEO of a Fortune 500 competing company and explain what you do and what you intend to do with your strategic plan, and the ability of a Fortune 500 company to react to that information would be effectively zero.

It's counterintuitive, but the more people who are aware of your strategic plan, the better chance you have of creating a successful company. So, tell everyone and anyone what your strategic plan is—your target entry market, the analysts, the investors, and especially your board and employees. You really have nothing to lose if you are unique and solving a critical need.

One example of a company that put their strategic plan out in public view is Tesla. From the very beginning, formatting explained how they would enter the electric vehicle (EV) market, first by putting their design on an existing car (they chose the Lotus Elise) to debug the technology, then they would manufacture their own very expensive and low-volume sedan to learn the manufacturing process. After that, they would introduce a high-volume, aggressively priced, mid-market car that would displace even non-EV competitors. Tesla's plan was published and obviously read by the CEOs of all the large car manufacturers and yet more than a decade later, those manufacturers were still struggling to develop a competitive response.

EXERCISE

CREATING A STRATEGIC PLAN

1. Explain how you will enter your entry market and make customers take notice.

2. Explain the specific approach that will propel you to success.

   - Make sure all your executives and your board have bought in and are fully supportive. You can even test that by having them repeat back to you the strategic plan like in a pop quiz in the classroom.

   - Remember to start every important meeting (all-hands, board, executive, etc.) by reminding everyone in attendance of the strategic plan's elements and what you're aiming to achieve through them.

YOUR OPERATING PLAN

The last part of transforming your vision into reality is going to be the foundation of your operating plan. Unlike the strategic plan, which is widely shared, the operating plan is an internally facing translation of your strategic plan into measurable objectives. For example, the operating plan for the Normandy invasion by the Allies might have been "Establish strong beach landings on at least three beaches by 4pm and have makeshift harbors deployed to allow resupply." For the financial services case previously mentioned, the operating plan might be "Win three midtier Midwestern banks by Q2, have them deployed by Q4 and be ready to address a wider market by next year." In the e-commerce example, your statement might be "Be fully certified by the appropriate bodies for privacy by Q1, identify and start Proof of Concept (POCs) in five pharmacies by Q2, and close first accounts by Q4." In the mobile devices case, you might decide on "Approach 100 insurance brokers individually in Q1, sign up at least 20 by Q2, have 300 daily users by Q3 with a net promoter score (NPS) over 70."

At this point in the process, a startup should be setting the overall direction and merely quantifying the targets in their strategic plan. The high-level goals should be such that succeeding at them would effectively get you to the objectives in your strategic plan. We will discuss later how to translate this general operating plan into a detailed plan that will serve to operate the company on a quarterly granularity. Although the operating plan you develop at this stage is not as detailed as what it will eventually become, this initial version is critical to some specific stakeholders. For example, your investors will want to know that the objectives you set will allow you to raise your next round of funding. Your employees will want to know that the entire company is aligned toward accomplishing those important milestones and that the future is bright if they're achieved. It's important to have an operating plan at this early stage that satisfies at least those two important groups of stakeholders.

Your startup is not solving a problem generically for anybody in the world. You are addressing a very specific need for the people with that specific problem. The process described here is very difficult to do and a lot of people, when I share this exercise, will tell me, "Nobody on our team wants to do this because it's really hard." Yes, it is hard, and you need to acknowledge that it is hard—and also recognize that your market entry stage behavior is not like anything else in the life of a company that scales. But in doing this exercise, you will establish the quantifiable foundations for your market entry, and that work is incredibly difficult. It is also incredibly valuable.

EXERCISE

CREATING AN OPERATING PLAN

1. Define what are the key elements that you will execute in your operating plans that let you achieve your strategic plan.

2. What objectives will you set yourself that will enable you to win?

3. Validate how executing these objectives will ensure you a winning market entry.

THE MARKET ENTRY STRATEGY ON *THE MARKET ENTRY STRATEGY* BOOK

As an example that is simple to understand, here are what I came up with for the vision, mission, strategic plan, and operating plan for writing this book and launching it:

Vision

- There is an increase in the number of startups and they struggle to get to market.
- The failure mechanisms are centered around differentiation, focus, and execution.

Mission

- To provide the startup community with a reference handbook to lead them through market entry stage.
- To improve the outcome of many startups by providing a disciplined and thoughtful approach.

Strategic Plan

- Write a book based on the materials of the market entry strategy seminars.
- Enter the market by focusing on the attendees of the market entry seminar, then leverage a publisher network to reach a wider audience, having obtained readership and reviews.

Operating Plan

- Work with book coach to deliver a fully baked draft of the book by year end.
- Have initial reader feedback by Q1.
- Present the book and refine it with publisher, publish by Q2.
- Reach out to the seminar attendees and encourage them to leave reviews.
- Publish widely by Q4 of that year.

PUTTING IT ALL IN ONE SENTENCE

Positioning and messaging for market entry require all four steps outlined here: vision, mission, strategic plan, and operating plan. By completing work around these four elements you will now have full alignment on each. The ultimate litmus test though is to synthesize and summarize all of your thinking into *one single sentence*. Although there are many variations on this theme, the sentence I have seen used and I like the best is the following:

Company _____ [your startup]
Provides _____ [your unique business benefits]
For _____ [your entry market customer]
Who need _____ [the critical need you are addressing]
As opposed to _____ [list every competitor and why they
are unable to solve this problem for these customers]

Notice how this sentence captures the key elements of what we have spent this whole chapter working through:

- The solution and business benefits that your product and technology enable.

- The specific entry market customer segment you intend to serve.

- The critical need you have chosen to spotlight to compete in the entry market.

- The uniqueness of your solution compared to every competitor you can name.

You should be exhaustive in the last part of the sentence, in naming each and every possible direct and indirect competitor and explaining why specifically they can't do what you can do. This one-sentence exercise is going to prove whether or not you have nailed your Market Entry Strategy and provide the foundation of all your messaging. Again, it is best to make this an off-site team exercise because, by construction, this sentence is one of the most challenging things a startup can do. Trust me, it's hard, because you are now forcing the management team and the entire company to coalesce on one thing and one thing only. But if you can write this one sentence, it's a beautiful thing because everyone will be on the same page. Your marketing materials will be consistent, your messages inside and outside of the company will be the same. This sentence will be the tagline on your website, it is the background of your booth at a trade show, it's the 30-second elevator pitch, it is the most important slide in your deck. It is the sentence you will train your salespeople on and your sales development representatives to use as precisely as possible. It will be on your customer qualification sheet and in your competitive objection playbooks. You will be amazed by the benefits of defining your startup in one sentence because now you are in sync with your materials, in sync with your talk track, and in sync with the way you will set up your technical engagements.

The Market Entry Strategy **book in one sentence:**

The Market Entry Strategy
Delivers a comprehensive and disciplined methodology
For startups at market entry stage
That struggle with execution, positioning, and focus

As opposed to:

Crossing the Chasm, which relies on an
early adopter methodology
Lean Startup, which emphasizes fast random iterations
Startups just doing by the seat of their pants
and not having a plan of record
Biased and superficial market feedback
based on investor connections

Practice saying your sentence over and over again and pay attention to how it *feels*. If you are making an unnecessary effort, if you are struggling with the words, or if you feel like your body is rejecting it, pay attention to that. These are the telltale signs that the sentence is still not right.

Also look at the sentence and objectively analyze if your company—and only your company—can use it. If your sentence could have a competitor name replace yours, it is not right. Think if your target customers will recognize themselves—and themselves only—as having the problem that you uniquely solve. Think about whether your target customers would wholeheartedly agree that your definition of the problem describes their situation and your unique solution resolves it. If they agree that you have defined their critical need but don't believe you can solve it, or they challenge you to prove it, *you have won this part of the battle.* As we will see later, you are in an ideal position to execute a technical proof of value completely aligned with your spotlight.

Every salesperson, every person representing the company, will need to learn this sentence. It will be hard enough to maintain consistency in using the one sentence throughout the company if your management team is not 100 percent committed to use it without any variation. If there is not this absolute commitment then one day you will listen to one of your employees pitch and you won't recognize your own company in their words.

EXERCISE

THE SINGLE MOST IMPORTANT SENTENCE

1. Develop a draft of your one sentence.

2. Chisel it again and again until it is the shortest possible. Try replacing words with synonyms and see which one works better.

3. Make sure it is unique, that no competitor could say the same thing.

4. Imagine walking into a trade show and seeing it on your booth or reading it on your website. Are you ready to commit to it?

5. Have each team member say it aloud, see how they "feel" it, see how it resonates with you.

Creating one sentence for your startup matters—a lot. It matters be-cause once you create it, you can't change it every few days or weeks or you will fall victim to the lack of spotlighting. You will have to polish it and stick with it at least for twelve to eighteen months to have any chance of anyone else other than you remember what your company does, which problem it solves and for whom, and why only you can make these claims.

KEY POINTS

- Your strategic plan details the key objectives that will lead you to succeed in your market entry execution.

- The strategic plan should include key strategic actions, customer focus, and how winning them leads to market leadership.

- The operating plan is an internally facing translation of your strate-gic plan into specific, measurable, timed objectives.

- The vision, mission, strategic plan, and operating plan are all aligned and derive from one another.

- The "one sentence" will summarize your complete positioning and messaging and will be the foundation to all your communications.

10

Market Leader

"The person who follows the crowd
will usually go no further than the crowd.
The person who walks alone is likely to
find himself in places no one has ever seen before."

—ALBERT EINSTEIN, theoretical physicist, Nobel Prize recipient

If you followed the Market Entry Strategy you will have identified a key benefit, for a target customer, in a specific sliver of the market that will be your entry market. Although getting to this point is a lot of work, it's worth it because it delivers one of the most valuable assets a startup can get in that you are the de facto market leader. You might dismiss that fact as secondary since the entry market might be very small because you reduced it by using the market entry principles outlined above. But, by construction, you are indeed the leader in that market because you have chosen it such that you have a unique benefit. So, when you manage to penetrate the customers in your entry market, they will see your startup as the market leader in that market. Although you might think of market leadership before you have made one significant sale as something made up or untrue, it is fundamentally different. You are not falsely claiming that you're a leader, you are in fact a leader of that specific market, uniquely solving a critical need for those specific customers. And the fundamental differentiation that you have worked so hard to articulate is the key to that leadership.

Even though it is not a B2B example, I like to refer to Subaru as a market leader in a sliver of the market because people are familiar with the brand. Subaru is a small car manufacturer that cannot afford the technological investments that its larger competitors are making. Its flat-four engine architecture is outdated and the performance of its cars on many metrics is undifferentiated. But Subaru has focused on one sliver of the market: the outdoorsy hip pet owners. You will not see any Subaru advertising in the general media and their messages only speak to their target market about using their cars to go into nature or carry their pets along with them. And so, in that sliver of the market, Subaru has managed to become the market leader!

As the market leader, one extraordinarily important benefit you get is that you control the narrative. You get to articulate what this sliver of a market needs for this specific problem that only you can solve. As we have seen in the previous chapter, you describe your vision for how this market is coping with the challenge you are tackling and how it ought to be solved. As the leader, you have defined the problem and chosen the words to describe it since there is no prior state of the art, no prior competitors or solutions in your entry market. When you educate the customers and the market watchers on your vision and your nomenclature, they will adopt it because you are in new territory. You control the vocabulary that describes your market that your product is positioned to lead. Once the words have been decided, your possible future competitors will find themselves in a battle that has been set to your advantage since you defined everything to maximize your differentiation and spotlight your unique benefits. They will have to play in your game and can only challenge you on your terms. In other words, you define your competitive landscape and you get to position yourself as the leader and you will consequently deposition your competitors who have to play within the framework you have defined. You also get enormous power with the analysts and influencers in the market because you get to call things the way you decide that they're being called. And everybody will start using those same words that you have chosen because, in this sliver of the market, you are the leader.

Because you are the market leader, you also decide how to define the customer product requirements and what critical components a solution to this problem must include. For example, if an accuracy of 99.7 percent is needed to achieve the relevant results or if a fusion of seventeen data sources that you explicitly name is required to solve this problem, you get to set that bar. If a system must be fully automated or remotely operated to solve for the particular configuration of your entry market, you get to define it to be so. If you are able to list five critical elements that make your solution unique in your market entry, then your job will be to prove that anyone that wants to solve this problem must have at least the five that you have defined. For example, because you define both the problem and the solution, you are able to say, "This is what the product to solve this problem looks like. Unless you can do 1, 2, 3, 4, 5, you really can't solve this problem." Of course, you are doing it for yourself because only you can do 1, 2, 3, 4, 5, but then you get to say to customers and to the market, "Competitor X is great, they're amazing. They do 1, 2, 3, and 6, 7, 8, but they don't do 4 and 5, and in this market, the requirement is 1, 2, 3, 4, 5."

The epiphany of the Market Entry Strategy, and the opposite of iterating, experimenting, and leaving decisions to chance, is that the process is *entirely in your hands*. You decided what to put under the spotlight, and you chose the matching entry market where you can become the number one leader in the category. I haven't said anything that is not under your control and so you can create your own destiny. As a startup, would you rather leave your entry market to chance or sit down in a room with your team and think this through?

As a startup executive team, you should want to retain control and decide for yourself what customers and markets you want to participate in. You may be wrong—not everybody gets it right—but would you rather have it be your choice and think through the entry market prospects that you have chosen, or leave it to chance? Very often we see startups talking to whoever is willing to listen to them. That is particularly acute in environments where the entrepreneurial ecosystem is well-connected, where everybody knows everybody else and people talk to whoever they can get to. But whoever you can get to may not be the people you want to get to. The question you need to ask and answer is, "Do the people that I'm talking to belong to my chosen entry market and are they viewing my startup as the market leader?"

Early stage investors will ask startups why they picked the customers they did—what made them be the customers that you wanted to talk to? What makes you the leader in the markets that seem to be defined by the logos that you highlight? If those logos are all over the place, the immediate reaction will be, you are not leader of anything. You are dabbling in a bunch of things, but you are not leading anything. An investor would be very worried about how you are going to scale and spend the money that you are given if you undertake such a broad attack. Going back to the Normandy example, you are effectively claiming that you are going to conquer those beaches, but you simultaneously have ships in Brittany, troops landing in Bordeaux, and you are sending commandos on the Belgian shores. That's not going to work because you have to put all your efforts into one entry point. To make it to the beach you need to concentrate your forces.

You may be thinking, so what? Who wants to be the leader of a very small market? Why be a market leader? Some people may think that sometimes it's good to be a fast follower instead, to penetrate the market when there are already participants. This is incorrect and even if you want to be a fast follower, then you would want to be different along some parameters of the existing offering. In that differentiated approach, you would want to be a leader. Let me detail the many advantages that a market leadership position affords you, in addition to controlling the narrative.

REDUCED COST OF SALES

The absolute number one benefit of being market leader is that it leads to reduced cost of sales, and that is a huge benefit. At the beginning of the life of your company, you have very few sales resources—if any—and the market is very large compared to your capabilities. A startup is not going to be able to tell the whole world about their product directly, it is not going to have enough salespeople who are going to find out that somebody in some corner of the United States is looking for the solution you provide. A startup won't have enough marketing dollars to make sure that whenever people search something on Google in their category they'll find you. A startup cannot afford the popular generic keywords in their space because the obvious keywords are being auctioned at a really high price to people who have a lot more money than a startup. But, if you are the market leader, if you have made your name and your product synonymous with that specific problem in your entry market, then whoever is in that sliver of the market is more likely to call you. The search engines will identify you with the specific market queries and if you need to buy those long queries, the Google keywords that are specific to that sliver will be affordable and effective.

With very few exceptions, most customers within a given market segment are talking to each other. They may openly collaborate on certain topics, for example, cybersecurity, even though they are in cutthroat competition with each other. And even where they compete, they go to the same conferences and follow the same industry publications. If you are the leader in solving a critical problem in your entry market, it is your startup they will be talking to each other about. And you will reach a critical mass, that resonance of the word of mouth, much earlier than if your first customers belong to different market segments and don't talk to one another or are using your product to solve different problems. This singular perception about a startup is the result of spotlighting your differentiation, and as we have seen, it has a clear impact on your market entry efficiency.

The consequence of this free word-of-mouth marketing is a reduction of your cost of sales since you get people talking about you with no cash outlay. Very quickly, much quicker than if you had hired a large sales force or bought Google keywords, you will be in the position that people call you. And for a small startup, this is the most critical transformation in your life. Before that happens, you are sitting and thinking, "Somebody somewhere in Alabama or Minnesota or Oregon is buying something in this space, and they don't know I exist. I need to spend marketing dollars, and I need to hire salespeople, and I need to chase those deals that I don't even know exist because they don't know me." But if you are the leader in your entry market, the pivot to them knowing about you happens much earlier.

> I had a portfolio company that went after the sliver of the market in cybersecurity for medical devices whereas most players broadly went after all connected devices. Because they spotlighted a critical need in healthcare specifically and focused on the larger hospitals in the US, they quickly became market leaders in their chosen market. Much faster than most companies I worked with, there was a switch in the market perception and prospects started to call them—they didn't have to call the customer. People started buying even without a POC because they were the market leader and many of the people in their networks had bought the solution and successfully deployed it. They were invited to industry events, met with key customers in their target market, and were able to close large contracts with high sales efficiency. They achieved that market leader position in that specific vertical and none of the platform solutions in the broader market were able to challenge them.

As a market leader you will be invited to bake-offs and requests for proposals (RFPs). Once the entry-market-specific industry publications have you listed as the market leader, no decision-maker in their right mind will proceed with an evaluation of solutions without trying your technology, even if their first contact was with someone else. And when you arrive on site, they are predisposed to use your vocabulary and to set the criteria for their proof of value to largely follow the specific capabilities and benefits that you have set for the market.

And thus, market leaders benefit from a higher capital efficiency because their position positively impacts their cost of sales.

PREMIUM PRICING

In addition to increased capital efficiency, market leaders will also have a few additional tailwinds, such as the ability to command premium pricing. If your startup truly does something unique, something that is a critical need for your customers and that no other competitor can do, they are more likely to pay a premium for it. If you have a "me too" product or if you are only slightly better in a market where there are a lot of other solutions, customers will say, "Well, I like what you do but the other company does it for $10,000 and you are asking for $20,000. If you made it $9,000, then I'd buy from you; otherwise I'll buy from the other company." Evidently, your startup is not unique in that market because if the customer didn't have the option to buy from another company, then they would pay the $20,000 and not negotiate with you below the other company's price. But if you are the market leader solving a problem that only you can solve, the customer can't say, "Well, I have an offer from your competitor and it's half your price," because you can respond, "That's nice, but do they do 1, 2, 3, 4, 5? No, they only do 1, 2, 3, but you need 4 and 5 as well." If that customer still does purchase from your competitor, then apparently that customer wasn't properly qualified. You thought they needed your unique solution, but they didn't. So, if you qualify customers correctly, you should be able to maintain the premium price.

Being able to point to the benefits that only your startup can deliver is a critical weapon when faced with the pressure from the purchasing department that is used to squeeze the juice out of fledgling vendors. If your startup is unique and a market leader, and you can demonstrate to the purchasing people that only you can solve the specific problem, you have a lot more leverage because they know they can't walk away without your solution. If they attempt to create an artificial competitive bid, you will be confident that this is not a real alternative. If you are forced to sell your products on the basis of a lower price, recognize that it is a sign that you haven't been able to demonstrate unique benefits for these customers.

PREFERRED VENDOR STATUS

Another important benefit of being market leader is that customers will begin to view you as a trusted partner for the field in which you dominate. That trusted partner status translates into an openness to disclose their future projects and the associated challenges that you might be able to address. Having this inside information will allow you to coordinate your roadmap with theirs, further reinforcing your leadership.

Additionally, while a startup needs a product that is unique, differentiated, and one that solves a unique customer need to become a market leader, that only applies to your first product. Being a market leader in your entry market acts like a shield for your follow-on products. Follow-on products do not have the same requirement—the customer bought from your startup because your first product was truly unique. If the second product is not that unique or differentiated, it will be protected by your first product and the customer will be more likely to buy both from you since they need you as a vendor anyway.

Another benefit of preferred vendor status is what I refer to as forgiveness. This is something nearly everyone in tech has experienced: You put your product in the customer environment and something crashes. What does the customer do? They blame the new company, the startup that's the last one that installed their product. It does not matter whether that same day there was a cloud outage, it doesn't matter if they used the wrong version of the browser or they didn't install the last software patch, it is always the startup's fault. Whatever happens the day after you have installed your product has to be your fault because it couldn't possibly be anything else. But if you are truly unique, if you are really solving a problem that nobody else can solve, customers will more likely forgive you. If you are the market leader, they will be more compelled to want to continue working with you. If you are not that special, if you are not a market leader providing a unique product, then you are out, you are done, because there is too much friction and too much at stake to take the risk for a small benefit.

INDUSTRY VALIDATION

For startups in tech and especially in modern software, there really are no standalone products—everyone needs to interface and either receive some input from another system or write their output to another framework—they need a database, an API (application programming interface), or other things to integrate with. If you are the leader in that sliver of the market and you approach Salesforce or Palo Alto Networks or Amazon and you say, "I need access to your API" or "I need to work through the flow with you," if your startup is able to give them customers that validate your leadership position, then they will work with you and they will work with you first. And, if you need to be more persuasive to get access, your customers who definitely find your product critical to them will be able to supplement your efforts with their own pressure on those vendors. For the startups after you that show up to those same partners and ask the same questions, the response is likely to be "We are already working with the market leader. We're not jeopardizing our relationship by giving you the same access." A company like Salesforce or Amazon might still work with the second company in this sliver of a market, but not the third one. If you are not the leader, you might find yourself in a position where somebody says, "Well, the API is documented, so you need to figure it out on your own." The other embodiment of that predicament comes with the various application stores (app stores) where a vendor will display all the partners they work with. Being represented there without having a preferred status or a collaborative go-to-market is tantamount to being on the bottom shelf in a supermarket.

As an added benefit to being market leader, the more vendors cooperate with you, the more industry validation you benefit from. These vendors will have presentations in which your logo appears as a partner for your specific field and will expose them to your possible prospects. In many situations, you might be listed on their website and be listed among recommended vendors. They will also invite you to present at their user conferences and will make space available for you in their booths at trade shows and other events that you might not otherwise be able to afford. This will accelerate the resonance of your message. People will notice—whether it's venture capital investors or other people who are influential—they will see the startups making progress in the industry. The people in your industry see that it is happening, they see that you are making progress and they are thinking that you are going to be successful. You can quickly get to the point where your cost of sales collapses once you get industry validation because people speak to one another and you get free marketing.

The other kind of cooperation that is a competitive advantage for market leaders are the channel partners. If you are in a market where the channel in any form (reseller, integrators, consulting, or other firms) plays a role, the market leader is the one those channel partners want to work with. The top resellers carry a lot of weight and if you are the market leader, you should manage to partner with them before your competitors. Being market leader will accelerate your go-to-market and create obstacles for competitors.

LEADERS GET THE LION'S SHARE

Last but not least, if you are perceived by the industry as the leader in your entry market, you will also be recognized by the investing community. Investors prefer—by far—to invest in the market leader and this means you will get access to more capital and at a higher valuation than your competitors. This will in turn increase your competitive advantage as you will be able to invest more in your product roadmap, have more customers that will serve as references, and accelerate your sales motion. This circular reinforcement of your momentum can often snowball and have a dramatic impact on the outcome.

The key reason that investors prefer market leaders is that the value of an eventual exit—whether an IPO or an acquisition—goes disproportionally to the leader. As we have seen, market leaders get the attention of the acquirers early in their development. Market leaders can also more efficiently IPO as they will get more analyst coverage, have higher trading volume, and consequently gain a positive impact on their valuations. There are no exact numbers, but typically the leader ends up grabbing 70 percent of the total value created in the space and all the remaining competitors divide up the remaining 30 percent, with the crumbs decreasing even faster from second to third and so on.

KEY POINTS

- Market Leadership—even of a smaller market—is critical to your successful market entry because it lets you:
  - Control the narrative
  - Define product requirements
  - Enjoy reduced cost of sales
  - Receive preferred vendor status
  - Earn industry validation
  - Have the highest outcomes

PART III

EXECUTING YOUR MARKET ENTRY

11

Total Product Marketing

"There is a tremendous amount of craftsmanship between a great idea and a great product."

—STEVE JOBS, businessman and inventor, Apple cofounder

The Market Entry Strategy is focused on solving a specific problem for the customers in a startup's entry market but, regardless of how novel or innovative a startup's technology is, customers are mostly interested in the business benefits outlined in the company's mission statement. There is a tendency for founders to overestimate the importance of their technology or platform and, although in some cases you might run into technical decision-makers who recognize a breakthrough and understand the implications of a technology, do not count on that happening very often. Even in this case, the customer still wants the unique benefits you have promised, and will judge whether your product delivers them or not.

Often that myopic fascination of many founders on the technology behind their product can lead to a disconnect between the startup and potential customers. The Market Entry Strategy is designed to help startups understand how to focus on specific business benefits for their entry market, and importantly, the technology itself is only one attribute in delivering what those customers demand. To understand how customers view a new, innovative product in that context, it's best to think of *total product marketing*, which refers to everything that makes up the solution, including what happens before, during, and after a product is activated, as well as everything that contributes to delivering the benefit to customers.

THE BENEFIT IS THE PRODUCT

In the context of total product marketing, it's important to consider all the inputs that are needed before your product can be activated. For example, if you need access to a database from a global company like SAP, Salesforce, or Snowflake, then the access permissions and the interfaces needed to make your product work are part of your product. The friction and speed of execution of moving the data will be an integral part of your product because those factors affect the total experience needed to produce the benefit you claim. Similarly, anything that happens during or after your product is installed is also part of your product. If the system allows the customer to detect a problem, then the full remediation loop of that problem—even if you have nothing to do with the remediation loop—becomes part of the solution you are providing. There is little value in detecting a problem that cannot be remediated or detecting a problem where the cost of the remediation outweighs the benefit of doing so. If your product can sometimes produce false alerts, then the time spent weeding those out is also part of the customer experience. Your startup will be judged on the net throughput of only the valid answers you have provided, minus the time spent cleaning up the rest. The benefit needs to be easily quantifiable, and the total product marketing view will include whatever else is needed to deliver that benefit.

In terms of making your benefit quantifiable, think about finding a measurable endpoint and consider, for example, the framework offered by medical devices where a trial is conducted with a control group and then clinical outcomes are compared at an agreed endpoint down the road using a rigorous, statistically valid experiment. Although it may not be obvious to you nor even your customers what that endpoint is, a total product marketing perspective should force you to imagine the measurable endpoint and propose to the customer an experiment that would enable them to validate your benefit at that endpoint. A total product marketing approach will help you think through the many situations that impact whether a customer receives the benefits you promise.

> If you focus narrowly on your product, you may miss the real impact of your solution downstream. For example, we see many products with alerting systems that suffer from the issue of producing enormous amounts of alerts that needs to be triaged, managed, and dealt with in some way. While a product that provides alerts might be very helpful and a benefit to a customer, those alerts must be valid and actionable. If you have too many alerts, the customer just doesn't have the staff for addressing the underlying problem. Having 10,000 alerts is the same as having none. In addition, the customer needs to understand the cost and impact of the fix on their routine, or they may not want to execute the fixing loop at all.

EXERCISE

DEFINE YOUR PRODUCT

1. Describe the totality of the customer journey, from the original data source to the final data produced that delivers the benefits you claim.

2. What data and interfaces are required to execute your workflow? What permissions and integrations will your customers require?

3. What is the output of your system? Does it require further processing or integration to deliver your stated benefits?

4. Does your product deliver data that is meaningful only in the context of a remediation or some iterative loop?

5. Does your product generate false alerts or other data that needs to be further processed to deliver value?

For all of the above, quantify the friction and effort your customers will have to overcome to enjoy your stated benefits. You can use metrics such as time elapsed, manpower used, number of steps, dollars spent to achieve the outcome.

WATCHING YOUR USERS

Despite your best efforts to think through your product, you may not be able to anticipate how customers will use it and one way to overcome that obstacle, and gain insight into the real-world use of your product, is to carefully watch your users. It may seem like a trivial observation, but you will always learn from looking over the shoulder of your users without interfering with their workflow. Experience has proven time and time again that the sequence of actions you imagined users taking is often not the ones they actually perform. A best practice is to start observing what users do even before they use your product. What is the insertion point in the routine they had before your system was installed? At what point during their day—in their flow, in their thinking—do they think of your product, or need to run your product? What do users need before they can get going, and will they see enough value in the benefits you deliver to justify consciously or unconsciously making a detour from their usual routine to add your step? Countless products have failed because, even though they deliver measurable benefits, there was no clear insertion point. Customers simply cannot free up time or justify the distraction to obtain the benefit from a new product.

Once users are using your product, you will need to continue to observe them. Even the most minor actions, like where they move the mouse, what they look at while inside your product, are important. Do they pause while using your product? Do they ignore or fail to notice the information that you believe is critical? Although there are a lot of technologies available today that help you to collect this detailed user experience data—and those tools are invaluable—there is still tremendous value in your own intuitive understanding of how users use your product. Over time, through measured usage patterns and observation, you will begin to have a clearer understanding of your product and be able to discern which parts of your product are key value-creation deliverables and which are merely necessary steps. This deeper understanding of your product could then be used in your pre-sales efforts and could also be especially useful in post-sales customer service. For example, unless the key benefit creation points in your product are being touched at the expected cadence, you could still have a significant and growing usage and yet face unexpected churn. Understanding the user experience would help solve, or at least point to, issues of churn.

There was a company with a computer-assisted medical diagnosis system that provided powerful insights by analyzing all the images generated within the radiology department of a hospital. However, while the product was groundbreaking and offered many benefits to customers, there was no real insertion point. Who would look at those results? What would their specialty need to be? When in their routine would they find time to look at the results? What differentiated treatment would be decided based on those insights? Failing to understand the insertion point, and the consequences insertion points have for the customer, that company struggled to penetrate its target market and had a lackluster outcome.

Finally, you can gain valuable insight in both user experience and your product by observing what people do after using your product. Is the output fed into another business-critical process that depends on it from this point on? Are metrics created that have significant upper-management visibility? Does the output of your product become part of an essential auditable trail? The key element to look for is whether the benefit that you claim in your messaging does indeed prove itself in your entry market in a way that's measurable and undisputable. All those elements are also part of your total product marketing and should be understood and tracked no less than what you would consider your core product.

EXERCISE

DEFINE YOUR USERS

1. Describe what users actually do before/during/after using your product.

2. What sequence of actions do you expect them to take?

3. What reports do you expect them to look at?

4. Describe what problem the customers are trying to solve with your product.

5. Describe the exact use case that your product is solving in your entry market.

Observe your customers' actual usage against that blueprint and make sure they actually get to the benefits you claim to deliver.

USE-CASES CONCENTRATION

If you have followed the Market Entry Strategy so far, recall that an essential component is to solve a critical customer need in the market where you can most clearly and easily demonstrate your benefits. That means that if you have determined the answer, then you will likely have identified a narrow market and a small set of use cases, or perhaps just a single use case. You may be concerned about a single use case, but a good use case that fully demonstrates the business benefits of your product is actually a positive. For example, having a single use case can help you polish your flow and product marketing for these users and this use case. In addition, if the originating data are the same or similar, then the interfaces you need to build will be limited and the permissions you need to obtain to make your product work will be well understood by your startup and its customers. The benefit-demonstrating output will also be similar, and your knowledge of the industry-specific idiosyncrasies will begin to solidify. As a result, their jargon will become familiar to you and in further engagements you will be perceived as experts at solving this specific problem for this market segment. Customers will intuitively trust you and look for your knowledge of their problem, often accepting your POV structure without arguments or even giving up on this step entirely.

You'll also find that a focused market entry approach is something early-stage investors will notice in their due diligence. Sophisticated investors will be looking carefully at the use cases of your early customers to see if there are divergent trends among those cases. Why does that matter? If a startup's early use cases diverge, their product roadmap will not be able to satisfy the different needs and the startup will likely fall short of many customer expectations and fail.

I once attended a meeting of one of my companies in a narrowly defined market and although I understood the industry in general, I understood the specifics of this market vertical only superficially. The competitors of the company I was working with took a much broader entry market approach and on the surface, it looked like these competitor startups were tackling a bigger, more profitable market. But as I found myself in that meeting, barely able to understand the points being made under the pile of acronyms and the industry jargon, I realized that none of my startup's competitors could ever effectively compete simply because my startup's razor-sharp focus made them unassailable: their customer intimacy and ability to tailor a solution to this narrow market of their target customers catapulted them to preferred-startup status and they were the only one the customer wanted to work with.

Other challenges lurk if you stray beyond a focused market entry. For example, with a broad market entry you'll need to train your sales force to recognize the full set of possible use cases and the varied value propositions they need to convey to your prospect. It is difficult enough to train people on a single message and use case, so the idea of training them to recognize in an early qualification call what use case to propose to a customer is simply too large of a task. The same logic applies to other aspects of a startup's business: if you enter a broad set of use cases, it will be very difficult to create messaging, positioning, and other go-to-market materials needed to penetrate the market.

> I was involved in the acquisition of a company with a dozen very large customers. Prior to the acquisition, we carried out a lot of diligence and proved to ourselves that the customers were satisfied. A few months after the acquisition, it seemed to be a big success, until the executives walked into my office and asked that we discontinue certain elements of the product and notify the associated customers that we wouldn't serve them past a certain date. I was livid when I heard that suggestion, but the team correctly argued that we couldn't continue to serve the divergent use cases properly. Eventually, we would lose every single customer. That was a painful and expensive way to learn that a concentrated smaller set of use cases can be far better for a startup to manage.

It is tempting, when you get a call from a prospect, to pursue that company because of their interest, but potential customers who are not the use cases defined in your market entry will be more of a distraction than a help. By definition, a startup will not be able to satisfy many inbound customers unless they fit your market entry customer profile and use case—and you may have to temporarily walk away from the opportunity of a large purchase order. Why? Because if this inbound customer has a use case that differs from your main case, they will not be a satisfied customer and churn will be looming. Even if an inbound customer is satisfied, you have exposed yourself to a different roadmap than what you designed for your startup. You are letting the market randomly dictate your fate rather than choosing your fate; remember, you chose your benefits to spotlight your differentiation, your entry market, and your customers—the fundamental principle of the Market Entry Strategy. Holding the line on your market entry choices requires an incredible amount of discipline, but the benefits will be clear in hindsight. These recommendations obviously apply only during the market entry stage; once you have a solid market position you'll be able to redefine a new strategic plan that will support a wider set of opportunities.

CLEAR PRODUCT BOUNDARIES

The question of what should or shouldn't be inside your product at market entry is something that is continuously argued and debated by startup teams, with no clear-cut solution. Just as a startup is likely to be exposed to use-case drift, you'll recognize its mirror twin—often referred to as "feature creep," i.e., constantly adding functionality to please certain customers in an inconsistent manner. The Market Entry Strategy obviously cannot give you specific answers about what should or should not be included in your product, but it does provide a clear framework. Because you have chosen your entry market to be the one situation where you can most easily demonstrate your unique benefits, you should be able to have some conviction and clearly inform your point of view about the product boundaries. Everything you need to have in the product to provide those benefits must be in the product but once you have those elements, there is a sharp boundary to be maintained: anything that is not strictly necessary to demonstrate the benefits should be eliminated.

Of course, in every product category, there are table-stakes functionalities that are not unique to you, but they are necessary just to get to the benefits. For example, to make your product function properly you must read and process data, you must connect to certain systems, you must display some generic analytics. Those are fine things to have in a product but if you find yourself adding functions that expand beyond your unique benefits and in fact are me-too capabilities, you should resist the temptation to include them in your product because they are in fact not part of your benefits. When you add more elements to your product, you are broadening the list of competitors and, by definition, you are not that good at those other capabilities. You are lengthening the proof of value of your system too, because you are effectively asking the customer to test functionalities other than the ones strictly necessary to reach your stated benefits. And if you are weak compared to competition on these added functionalities (and especially existing products the customer may already be using), the customer will challenge you to justify why they should switch to your product. But, if you keep your boundary clear, you will have benefits that are complementary to the products your adjacent competitors offer to the customer and you won't be tempted to try to replace them. If you're hoping to steal that adjacent supplier's budget, remember that you have now made feature-parity the barrier you have to cross in a category you are not that great at. Perhaps your core benefit will be so great that the customer would settle for an inferior expanded capability, but this is a risky strategy.

> I was involved in a company that was clearly a market leader in its niche. It had grown to significant revenues but was looking at an adjacent, much larger market in which it had no competitive advantage and was not remotely close to feature parity. Although the company could have successfully sold the additional functionality as an add-on to its core differentiated product, the sales force quickly drifted into wanting to go into the larger market and having to prove their superiority in that adjacent category—which was futile—and the conversion rate plummeted.

Remember, too, the other side-effects that apply in the multiple use cases: you now have likely divergent product roadmaps, multiple messages to create and manage, and sales-force training to cope with this increased complexity. Experience has shown that the sales force is still likely to engage in a use case where the additional functionality is key to the customer and they care less about staying focused on your core benefit. And if that happens, you will enter into a costly bake-off that you most likely are going to lose. Even if you somehow manage to win in that wider feature set, you now have a customer who did not buy because of your differentiated benefit. And you'll be faced with fighting a permanent churn threat against competitors who have a better product for their own target market.

EXERCISE

DEFINE YOUR PRODUCT BOUNDARIES

1. Describe the functionality necessary for your product to demonstrate your unique benefits in your entry market.

2. Describe which functionality you will claim for uniqueness / superiority, and which functionalities are merely table stakes.

3. Describe the boundaries of your product, beyond which you will not attempt to go (and instead address through workflows and/ or integration with adjacent products).

4. Describe functionality that products have in adjacent spaces that you will not attempt to replicate nor get to feature parity.

PRICING AND PACKAGING

One element of total product marketing that often goes unnoticed by many startups is packaging, but it is vitally important because it impacts the customer end-to-end experience. Here, too, the Market Entry Strategy can provide guidance and help startups avoid costly mistakes. Quite simply, packaging includes how the product is separated into different components and how to charge for them. There are multiple ways to think about pricing, whether it is usage based or user based, whether the product is provided with or without a minimum fee, whether you charge a platform fee, or any number of other ways to slice and dice your product. Is some part of the product free forever or free for a limited time? Is there a component that you charge for separately? You may not know how you will package your product, but the important element is to think about it and analyze the business value of the benefits you are delivering and the alignment between the benefits and your pricing model. Connect your packaging to the benefits you deliver to your entry market and your strategic plan to see that they align. When you make that transition from entry market to broader market you might want to change your pricing and also consider how you'll manage the legacy pricing and agreements. If you think about that transition early in the life of your startup, it is not an insurmountable task, but if you wait until you enter the broader market to move to your long-term pricing strategy, you could have problems that are difficult to solve. So think through both your short-term and long-term packaging before arriving at your first use case.

As discussed earlier, one important consideration that startups should be aware of when it comes to pricing is that it is a poor strategy to win either on price or price structure. Why? If you have followed the Market Entry Strategy, your entry market customers should want to buy your product because you specifically solve a critical need, not because you are the most cost-effective solution. You ought to be the only solution for customers but if you find that you're failing to win when you price on par with your competitors, or that you win only because you are discounting more, that's a red flag to your positioning at least, or perhaps to your entire vision. Your competitors can always match your price if you become a sufficient threat. While being the low-cost solution is never a great startup strategy, as previously discussed, there is one exception to competing on price. That one exception is if, *and only if*, the product approach you have taken has such a significant material impact on the cost structure of the product that there's no possible way for your competition to match it. Note, though, that this advice does not apply to your competitor's profitability structure because, faced with a meaningful loss of market share, companies will adjust their pricing if needed, even at the cost of a temporarily lower margin.

EXERCISE

DEFINE YOUR PRODUCT PACKAGING

1. Describe how you will price your product and how you will keep track of the use / cost.

2. What is included in the basic product? What is free? What will be an additional cost?

3. If you have any kind of freemium (i.e., a version of the product is free and another that is paid), where will the payment boundary be (such as time or function)? Why do you think this will make it easier to demonstrate your unique benefits?

4. Knowing your pricing and your cost of goods, estimate what your gross margin will likely be at market entry stage and also in the long-term financial model.

THIRD-PARTY WORKFLOWS

In today's environments, there is almost no product that exists in isolation as a standalone product. Nearly every product connects to something else, like other software, or a database, a public repository, a platform that validates access and identity, or many other tools and systems. This list is not exhaustive and of course each startup will have multiple connections that need to be made to get their product to work properly. If these third-party elements are part of your workflow, they are also part of your total product. Customers will expect that you've worked out the kinks and issues with any third-party elements, whether those issues are technical, legal, or operational. Customers will expect you to remove any obstacles from their ability to enjoy the benefits you touted. If anything goes wrong, if anything breaks down, if there is any disconnect, customers will hold your startup responsible, not the existing other product you are connected to, and certainly not themselves for not having secured that connection.

Keep in mind that the Market Entry Strategy can help you with critical third-party relationships in narrowing the scope of those connections and picking the limited set only needed in your entry market. Keeping only the third-party connections needed for the specific benefits you are trying to deliver makes this more manageable. You are more likely to get a proficient and trouble-free end-to-end experience with a small subset of integrations. The important point is to recognize that these integrations are also part of your total product and that they need to be addressed before you go to customers.

EXERCISE

WORKFLOWS AND INTEGRATIONS

1. Describe all the integrations needed or desired by the customer to be able to use your product and demonstrate your unique benefits.

2. Show the plan you have to obtain or facilitate your customer's obtaining these integrations.

3. What legal framework will you use to obtain integrations?

4. What permissions or licenses are required?

5. What integrations will you decide to postpone and exclude from your total product during your market entry phase?

SUCCESS STORIES

Surprisingly perhaps, success stories are part of your total product marketing. How can a startup have success stories if it does not have any customers? As strange as it sounds, I strongly recommend that startups create their success stories before they have their first customer and even consider writing them *before* they start developing their product. And the way you do that is to write a success story—actually write down what success looks like from a hypothetical customer's point of view. The success story you write will be aspirational and motivational; it is much easier to know what you are trying to accomplish if you know what success looks like before you start marching toward it.

You might think of this approach as pushing the boundaries of truthfulness, of saying something that may never come true, but it is fundamentally different because you are not putting your success story with real companies on your website or marketing materials. This is an internal, aspirational exercise; the success story you are writing is an early version of what it would look like if your product delivered its benefits. A good success story will fully characterize what market you are in, what problems your customers have, and what benefits you deliver—it is the translation in the voice of the customer of your vision, mission, and the unique benefits you will provide in your entry market. The success story you write should represent how you have convinced a customer that is representative of your entry segment not only that they see the same problem but also demonstrates that you can uniquely address a critical need. Your story ought to showcase how your startup has been able to demonstrate the business benefits you claimed and represent an example for other customers as well as your own stakeholders—investors, analysts, and especially your own employees. Your success story should include how you inserted yourself into the customer's workflow and how your product was used. Success stories are an essential part of your product because they guide everyone inside and outside of your startup to understand what success looks like; they also model the route ahead for your trailblazing customers.

As odd as it sounds, writing success stories before you have customers—before you even have a product—is not just a funny exercise. It has very real implications for your ability to project your vision to the outside world and is definitely an integral part of the total product marketing.

EXERCISE

WRITE YOUR SUCCESS STORY

Imagine your first three customers each write their own success story. For each customer, write a success story that includes the following elements:

- Describe the customer in your entry market.
- Describe the business benefits that customer has observed.
- Describe how that customer has been able to prove those benefits.
- Describe how the customer's setup changes before and after your product.

KEY POINTS

- The term "total product marketing" refers to everything needed to deliver your business benefits, including what happens before, during, and after your product is activated.

- Fewer use cases that clearly demonstrate the benefits are superior to having multiple use cases that are less focused. Avoid use-case drift!

- Product boundaries should only include those elements needed to demonstrate your benefits; everything else should be eliminated. Avoid feature creep!

- Integrations are part of total product marketing and need to be addressed before you approach customers.

- Your success stories are part of your product because they define what success looks like for your startup and your market entry customers.

12

Competition

*"The most meaningful way to
differentiate your company from your competition
is to do an outstanding job with information."*

—BILL GATES, businessman and Microsoft cofounder

The idea behind the Market Entry Strategy is that startup success is tightly connected to how it defines an entry market where it will have unique benefits that no other competitor can claim to provide. If a startup can define an entry market with those characteristics, it will gain the amazing advantages described above. Underpinning those concepts, it is obvious that for a startup to succeed in that, it needs to have deep and continuous knowledge about the competition. A startup will need to know its competitors' capabilities to be certain of its uniqueness and will need to have clarity on the next steps those competitors are likely to take to be able to maintain this position. In this chapter we will dissect what this means for you and your organization.

CONTINUOUS COMPETITIVE ANALYSIS

Every company knows they need to understand their competition and to do that, they normally conduct a competitive analysis. Startups typically do a significant amount of research before and during the early stages. When they pitch to early-stage investors, they always have a nice chart in their presentation where they will necessarily be in the upper right-hand corner—often defining the axes they're measuring so they end up on the highest position. By itself, there's nothing wrong with that if the basis of competition in your market is what you measure yourself against. Remember that under the Market Entry Strategy, you get to pick your entry market and the benefits you put forward, but only as long as those benefits are relevant to the critical customer need. Investors too often see startups present graphs that are contrived to match some angle under which a startup looks good, but does not relate to how customers think. If you're not addressing an issue that's critical for the customer, you are in trouble before you even get started.

Even if a startup does an excellent job of presenting its unique benefits and customer needs, we often see that most companies do not revisit the competitive landscape until the next fundraising or until a customer tells them they've lost a comparative benchmark against a competitor they were unaware of. In recent years, venture capital has transformed from Silicon Valley to a global phenomenon and, although a startup team might hear about some competitors through word of mouth in their specific geography, today there could be several teams around the world going after the same entry market with similar technology and benefits. In most large markets, you might not meet competitors until both of you have already scaled and run into each other at international trade shows or much later at the same customers sites.

Ironically, if all startups used the Market Entry Strategy, the trend of finding competitors only after you have scaled would be reinforced because you might not have the same entry market and therefore won't overlap in the market until later in your growth. The Market Entry Strategy can also help startups react to the discovery of a potential competitor by correctly choosing a different entry market so that they can still become a market leader largely undisturbed. You'll eventually run into these competitors, but luckily by then both companies will be valuable and might coexist (or both might be able to find growth opportunities that continue to not overlap).

For the Market Entry Strategy to work well, startups need to always know everything possible about their competition—not just once in a while, and not just when they are starting out. It's important to monitor competitors in all directions continuously. The approach I've seen work best and which actually allows a startup with more tasks than available time to continuously monitor competition is to assign one or more competitors to each person on the executive team. Every vice president, regardless of function, should personally own monitoring one or more competitors. That task also applies to engineering, sales, product, marketing, and even to other executives (like the head of human resources or the head of finance). Everyone on the team should be involved in continuous competitive monitoring, and everyone who hears anything or learns anything about a particular competitor should share the relevant information with others on the executive team. Then, on a regular basis—probably monthly but certainly at least quarterly—hold a staff meeting dedicated to updating the competitive landscape where each executive represents the competitor they are monitoring. You can war game what could possibly happen now and in the future by having different executives create independent narratives and letting them argue the position of the competitor they "own." There are multiple benefits to sharing the added workload of continuous competitor monitoring; for example, you'll get a more balanced and complete view of your competitive landscape as it evolves. Also, because the team is collectively involved in this process and shares in the responsibility of the follow-through, you avoid any likely underestimation or finger-pointing.

The most important requirement in effectively monitoring competition is to be brutally honest about where the competition is compared to your differentiation and your unique value proposition. There is absolutely no gain in avoiding knowledge of the progress that competitors make or trying to sugarcoat the situation. You may think you are motivating your team longer by maintaining the appearance of being the "winning team," but you're really avoiding reality. Not only will that not last for very long, but in making up your own reality, you are being dishonest. I've seen startups increase their burn rate to fuel growth in a market they wanted to believe they dominated, only to see them crash and burn. Yes, you might temporarily see top-line increase but eventually—and usually quite quickly—this approach is not sustainable and startups will find themselves losing competitive benchmarks and having both customers and industry analysts dismiss their market leadership.

In the best possible outcome, a startup can perhaps reinvest in product differentiation and then return to the market. Note that often maintaining some focus on the initial market at this point is a mistake; the company would probably be better positioned for the future by laying off a significant portion of the team while they re-enter a new market from scratch. As they make this pivot they would probably have to find a different market—one with other benefits. The re-entry is the same as giving up your beach, going back to the ocean, and attempting a second landing. It is a lousy option for a startup but if you have to do that, you want to do it *before* you have a significant number of troops on that first beach. Otherwise, the lost time and resources you waste increases proportionally.

> I have been involved in a company that had an experienced founding team and went after a large market with a novel approach. Yet they quickly realized that they didn't answer a critical need and that proving the benefits involved a friction they couldn't overcome. They retreated, repurposed the technology to another market, came up with a new set of benefits for a totally different need, terminated the customers they had initially acquired, and changed the name of the company. This second landing succeeded and the company thrived because the retreat was complete and the re-entry was unencumbered by the company history.

Let me share an observation resulting from something I've seen time and time again: the team knows. As a CEO you might think that giving enthusiastic speeches about how superior your product is and how you are going to conquer your market is something others will believe, but they often are all too aware of the situation to buy into that. Yes, your team will clap at all-hands meetings and look convinced, but behind closed doors, they're seeing a different reality. They're facing your competitors on a daily basis and can see your startup's advantage dwindle; they see your sales engineers doing acrobatics to appear better than the product status allows; they see the engineers pulling all-nighters to try to close the holes and add some functions mid-benchmark. And so the people in your company know the reality and you, as an executive team, are losing your credibility because the reality you portray is not the one they see. They either think you are not being truthful with them, or worse, that you are actually unaware of the situation. Brutal honesty in your differentiation may not be pleasant in the short-term but it beats the alternative every time in the medium and the long term. If you lose your credibility as a truth-teller about the situation and the competition you are facing, you have lost your company.

EXERCISE

UNDERSTANDING YOUR COMPETITORS

1. How are you conducting competitive analysis?

2. Who is responsible for what competitive analysis in your organization? [list person and responsibility]

3. What competitors are you continuously observing? [list all] For each competitor, complete the following:

   - What is your true differentiation compared to this competitor?

   - How are you sustainably better than this competitor in your chosen bases of competition?

   - Are you meeting the table stakes with respect to this competitor?

4. Who/what is not a direct competitor but could provide an alternative to you?

TOTAL PRODUCT COMPETITIVE LANDSCAPE

The principle that your product is more than your technology and encompasses everything that happens before, during, and after customers use your product also applies to the competitive landscape. If, for example, your product needs connections to third-party platforms, those third-party platforms are part of your competitive landscape. If you need to provide your customers with the end-to-end experience and your competitors are seeking to do the same, then your competitor's ability to secure a privileged access to those other products means they have a competitive advantage over you. Those access points are part of your competitive landscape. Similar to the thinking in total product marketing, third parties are more likely to work with the market leader and reinforce the competitive advantage of that leader. The endorsement by third-party platforms either reinforces or diminishes your competitive position, depending on whether you are the market leader or not.

As discussed briefly in the chapter on market leadership, one other element that impacts a startup's competitive position is channel partners. In every industry there are a number of distributors that can make or break your startup, for example, value added resellers (VARs) and system integrators (SIs) that have privileged access to key customers and are pretty much viewed as trendsetters, or even kingmakers. Channel players, even more so than third-party providers, do not work with directly competing companies—they place one bet with one company. If you manage to secure a channel partner, your competitors won't have access to them and of course, if your competitors secure a channel partner you will be left out. Make sure to include channel partners in your total competitive analysis or you might be surprised; if you do not have a channel partner, you'll be hampered in your efforts to ramp up sales no less than if your technology advantage dwindles.

One last factor to include in your competitive landscape is analysts. These professionals can have a real impact on customer behavior. It's important to create a solid relationship with them, especially if you have a "new to the world" product. In the early part of your market entry, you will likely be in a position to educate them and although the best analysts usually have their finger on the pulse of the market, most of their understanding comes from meeting vendors and talking to customers. They will ultimately have a strong influence on your market penetration but in the early days, you have to work for them; you need to explain again and again your vision, mission, and your strategic plan. They'll keep track of your progress in the market and if you succeed in reaching market leadership, they'll effectively crown you and accelerate the second phase of your market attack. Their coverage or lack thereof is one more aspect of your competitive landscape you must track.

EXERCISE

COMPETITIVE LANDSCAPE

1. What are the critical third-party flows and where are you versus your competitors? [list all]

2. What are the critical channels and where are you versus your competitors? [list all]

3. Who are the analysts covering your entry market and how are you going to educate them to see your view of the entry market's critical needs and the benefits?

PROSPECTIVE OUTLOOK

Startups typically do a good job in their competitive analysis when they focus on a specific competitor or group of competitors, but there are things many teams overlook that, if addressed, could make their analysis stronger. For example, teams will routinely have a competitive analysis that begins at the present time and then maps out how their own situation will improve as they develop the additional capabilities on their roadmap. In doing that, they often compare their competitive position in the future to what their competitors are doing today, which is a false comparison. Obviously, competitors also have a roadmap and you should assume (and model) that they are equally capable as your startup, and assume also that they are equally likely to be funded or even receive better funding than you. It is best to role-play what your competitor's roadmap is likely to be in the future and then do the competitive analysis across multiple time points: today, six, twelve, and twenty-four months ahead. This analysis will assume that both you and your competitors will execute on their own roadmaps, a more realistic approach than thinking that only your startup will execute and everyone else is static.

The second element that will strengthen your prospective competitive analysis is to always pay more attention to startups that are emerging, not just to existing competitors. There will be some competitors you know about firsthand—you've run across them at trade shows, or customers have talked about them, or you have learned about them in the media—but it's equally important to find the competitors you do not know firsthand. A best practice is to do deep research in other geographies, and also look at adjacencies where a competitor is not currently in your market, but their logical product expansion puts them on a collision course with you. It's especially important to identify competing startups that began their market entry ahead of you, executed well (perhaps following the Market Entry Strategy), and now have a solid beach from which to expand and attack your startup. They will be attacking you as part of their expansion rather than as part of their first launch, and that makes them more formidable.

As part of continuously looking ahead, a startup should also understand large competitors, which is easier since their position is better known, and their evolution can be extrapolated. Most large organizations are slow-moving, big ships with a clear direction and less ability to make significant shifts. Large organizations are more predictable, and you can better plot your current and future differentiation with more certainty. It is never wise to ignore the impact of large organizations on a startup's competitive position, but they are seldom a mortal and sudden danger.

In aggregate, large competitors are dangerous but less than a startup in stealth mode can be. Startups you do not know anything about that emerge seemingly out of nowhere can create an upheaval in your entry market and a significant hindrance to your progress. I don't think I've ever lost a company to a move by a large competitor, but many venture investors have seen their hopes dashed because a nimble competing startup emerged and dominated their market.

MARKET DIRECTIONS AND MARKET SHIFTS

The third and most important element of the prospective competitive landscape is to include your ideas and analysis of market directions and market shifts. Market directions are usually known and can often be determined from the current situation without rigorous research. For example, a market direction might be that cloud computing will continue to become less expensive, or that customer data volumes will continue to expand quadratically, or that image resolutions will increase regularly. These are examples of market directions that you can work with and assume that they will evolve. Your competitive analysis should definitely include something on market direction, and you ought to be able to explain if the market direction reinforces or weakens your position—and what your startup will do about it.

In contrast to market directions, market shifts are much less obvious and more nuanced. Sometimes, for reasons that can be foreseen (or more often for unknown reasons), the market makes a sudden change in direction. When that happens, any assumptions you made, any ecosystem you were counting on, any customer behaviors you thought were consistent, can change in unpredictable ways and render your position much more tenuous or sometimes untenable. Market shifts can impact every industry. For example, there were massive investments made by most car manufacturers to improve the efficiency of diesel engines because diesels were thought to be less polluting, but when it came to light that Volkswagen and others falsified the data and that diesel was more polluting and affecting people's health, the market shifted to electric vehicles and most research and investments were redirected immediately. In the cloud infrastructure world, Docker emerged as the absolute leader in container software and was worth billions of dollars. But then Google open-sourced and propelled Kubernetes into the market. Within months Kubernetes became the de facto standard and Docker was only the shadow of what it was.

I experienced a market shift situation when I was involved in a company building on top of OpenStack, the open-source competitor for cloud infrastructure. OpenStack received many endorsements from would-be cloud infra-structure players but the system always had limitations and was never widely deployed; eventually that tech-nology became irrelevant. The company I was involved with could not recover from the market shift away from OpenStack. We did manage to sell the company for its technology and had a decent outcome, but were never able to overcome the market shift.

My advice is to stay alert on both market directions and market shifts because they can have a dramatic impact on even the best of plans.

EXERCISE

COMPETITIVE FUTURE

1. What are the long-term trends in your market direction?

2. What are the hot new topics you are tracking?

3. Imagine other brewing trends that could emerge to be significant.

4. What startup have you heard about but aren't sure you know everything about?

5. What startup is currently adjacent but could expand to become competitive?

6. What large companies are you tracking and can extrapolate their product evolution in your space?

VISUALIZING YOUR COMPETITIVE POSITION

If you've been following and applying the principles in the Market Entry Strategy, you will have analyzed your startup's DNA and chosen the entry market where you can most easily demonstrate your unique benefits. Your entry market is where you have developed customer intimacy, and with that knowledge you also understand the different bases of competition that your customer uses in deciding to adopt new products in your space. One of the best things you can do with this information is to display it visually, and the spider chart (also known as radar chart) is a convenient way to capture this information. The chart, as the name suggests, conveniently plots the different bases of competition as if they were the main threads of a web. You can use this format as the basis for competitive discussions with your team by creating your own spider chart and filling it with your data.

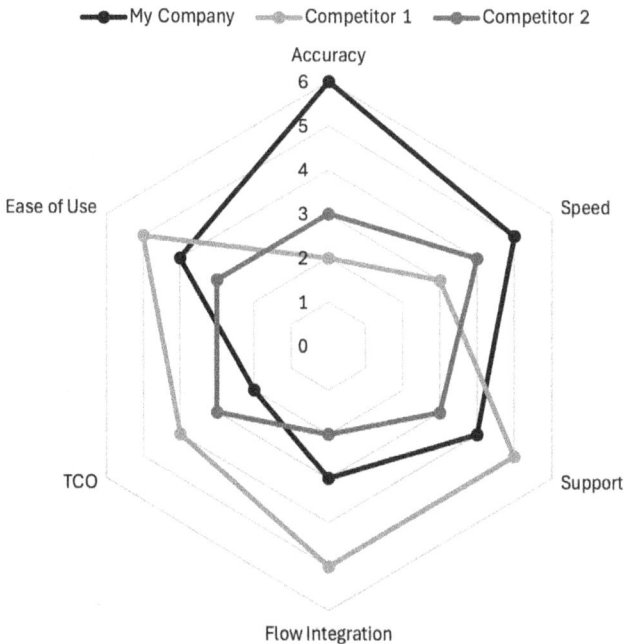

Competitive Radar Chart

Each basis of competition is on one axis, and in this particular example, the company has determined that the bases of competition in its market are price, cost of ownership, flow integration, ease of use, time to results, support quality, accuracy, and speed. Those are purely for illustration purposes and what you put into your own chart, what you determine are the bases of competition, are based on your own customer intimacy, your own deep understanding of your market, and what you've learned about how your target customers make decisions.

The axes are numbered and the values on each axis represent the strength of your competitive position. Then you merely plot how well you're doing against each axis and also how your key competitors are doing along those same axes. The spider chart should of course include the axes you've chosen to compete in your entry market, and your unique benefits should be visually obvious. Clearly, a spider chart provides another litmus test of your Market Entry Strategy and if you're surprised by the results, then take it as a sign to reread this book and begin again. Needless to say, brutal honesty is essential for this to be of any value. If you inflate your own scores and deflate the scores of your competitors, you'll obviously have an inaccurate spider chart, perhaps creating a false sense of security for your startup. You should of course also constantly update the chart based on information you obtain from your continuous competitive analysis and make sure the executives responsible for the tracking of each competitor agree with your portrayal. You can also use this chart as a basis for your sales process, training your team to understand each component of the spider chart—again the bases of competition—and be able to clearly speak with customers about your differentiation. It can also be used as a way to qualify prospects; if potential customers don't agree that the bases of competition on your chart are important to them, they won't care about your competitive advantages.

One last important point about the spider chart is how to respond if competitors improve dramatically on one of the axes you chose to be your competitive advantage. What if, using the example above, a competitor goes from "2" to "10" on speed, and speed is one of the differentiators you chose? The natural reaction of a startup is to counterattack and reduce the advantage the competitor has on other axes—to give up some ground on the speed differentiator and focus on catching up on the other bases of competition. Unfortunately, this natural reaction is critically wrong. If you decide to shore up your position along the other axes of competition, you will eventually find yourself with no differentiation at all and your Market Entry Strategy will fall apart. Instead, my recommendation is to double down on the axes you've chosen and try to increase again the differentiation against your competitors in this respect. In the example above, rather than concede the speed differentiation, you would increase your investment and resources to retain your speed advantage. Obviously, you have to achieve the minimum table stakes in the other axes, but discipline and restraint are needed to overcome the initial impulse and continue to maintain your investment in the key differentiation that matters most for your entry market customers. Avoid falling into the trap expressed via this statement: "Given enough money and enough time, all the participants in a given market have the same product."

I was involved with a company that had developed a simulator that was significantly faster and easier to use than the established technology for a specific set of use cases for which those characteristics were critical. We were dominating our market until a new competitor launched a product that was even faster and much easier to use. Instead of trying to broaden our benefits and concede our advantage in our entry market use cases, we started a complete rewrite of our software and reclaimed the speed and ease of use. The new competitor was soon relinquished to a distant second position.

EXERCISE

CREATE YOUR SPIDER CHART

1. Enumerate and validate that you understand the right bases of competition.

2. Make sure you accurately track key competitors.

3. Plot your position in the market and validate that it matches your positioning and your focus on your unique benefits.

4. Evaluate your product roadmap's impact on your future competitive advantage on the axes you chose.

KEY POINTS

▨ For the Market Entry Strategy to work well, you must be brutally honest about your competitors' capabilities.

▨ Competitive analysis is continuous and should involve established and startup companies, third-party providers, channel partners, and industry analysts.

▨ The competitive analysis must include market directions and market shifts.

▨ A radar or spider chart is a good tool to help visualize your competitive landscape.

▨ It is critical to double down and keep investing in your chosen differentiation rather than seeking parity with competitors.

13

The Other Side of Focus

*"The successful warrior is
the average man with laser-like focus."*

—BRUCE LEE, martial artist and actor

Every entrepreneur and nearly every manager will claim—without a doubt—that they are focused, that building their business, or refining their product, or whatever is relevant for their strategy, is the only thing that matters, and they are 100 percent focused on it. In fact, "focused" is one of the most frequently used words that comes up in every presentation to investors. The truth is that entrepreneurs really believe in their core that they are focused and are constantly making decisions that keep them moving toward their North Star. And yet in just about every presentation, there will be an incongruous element that suggests that, while they believe they are focused, they are actually not truly focused. When questioned about instances where focus seems to be an issue, an entrepreneur will often say that "this is the only exception" or "it is part of the strategy to do two things at the same time." In the dual-strategy case, their argument will often be that their technology can do several things, whether it is a parallel go-to-market, or two disjoint entry markets, or several packaging options for different use cases, or targeting multiple geographies. The list is endless of what entrepreneurs come up with that leads to less focus, and they have almost always an answer for why a startup is pursuing more than one thing.

I'm not judging people—far from it. The reason that focus is so difficult is that we are all humans and the amount of uncertainty in the startup journey is overwhelming. The number of forks-in-the-road, side avenues, or possible opportunities that a startup can take provides everyone with an infinitely large set of decisions. And because a startup at the market entry stage is so early, it is easy for entrepreneurs to get lost in trying to get their head around the longer-term implications of each of those decisions. An understandable pressure, when having to make all these decisions, is to keep several options open, to not focus. There is nothing harder for a startup to do than to say "no," to cut off that branch of the decision tree forever because maybe, with more time or money, that would have been the better path. To really focus is exactly that—it is to accept that with insufficient information, you need to make a decision that will impact the company forever, as opposed to constantly arbitrage and try going down several paths because maybe with more information you could make a better decision. The definition of *the other side of focus* is to understand what you will *not* do. The Market Entry Strategy is a methodology a startup can use to at least examine the decisions you face with a framework that can guide those decisions or, at the very least, serve as a mirror to the principles you can use to help make better decisions.

THE "NO" LIST

In the 1990s there was a best-selling business book by James C. Collins, *Good to Great*, that advocated a "stop doing list": things people in large organizations should stop doing—as a countermeasure to the pervasive "to do list." For startups, a similar problem emerges because people in startups are faced with many, many opportunities and different paths. Hence the startup as well needs to create a "no" list: an actual list of things they will say "no" to because those things are not part of their Market Entry Strategy. It sounds easy, but part of the difficulty of focusing is that as a startup, you are always facing a series of seemingly local decisions landing on your desk sequentially.

As a startup, you'll have to make early decisions that will prevent you from going down other paths in the future and you will not know until you get there the impact of a prior decision. No wonder you are overwhelmed when deciding on each fork in your road with incomplete information (such as knowing the eventual outcome of each path). One way to overcome the local and sequential decision problem, and one of my favorite antidotes to deal with the lack of focus, is to counsel startup teams to develop as good of a complete view of all the possible decisions they may have to face. In effect, you are trying to predict all the possible paths you might encounter. The goal is to imagine, a priori, all the decisions you might face and line them up concurrently. As a result, you have a more global perspective on how some decisions might stack up against others and you can more easily rank them and recognize which ones you will have to say "no" to and which ones are mutually exclusive of others.

The exercise I propose to startups that helps with focus is to have a brainstorming session with the team and in the most open-ended fashion you can think of, write out a list of all the possible decisions you might face. Do not limit your imagination; do not exclude anything because you want to approximate, as closely as possible, every possible path that your company might take. Think of a large company offering you a crazy sum of money to support their platform, or a foreign distributor with a ready customer for millions of dollars. How will you respond? Or think of an acquisition of another startup that would broaden your product portfolio, or a competitor executive quitting and wanting to join you. How will you react? Perhaps a strategic investor approaches you to inject a huge amount of capital with some strings attached, or a private equity fund wants to go down-market and invest in you. Create a big list of possible decisions your startup might face and how you might need to react when they appear.

Any single individual might only have a couple of ideas of what they would say "no" to, which is why doing this exercise as a team will lead to a more comprehensive list; one person's ideas will inspire others to think further. One of my favorite methods to generate that long list is to give everyone on the team a pack of Post-it notes and have each of them write out the items on their "no" list. Then, post them on a board and read them out loud—this almost always triggers more ideas. Keep following this process until your list becomes exhaustive and you are all tapped out with new ideas. Once you have this exhaustive, extensive list, the next step is to look at those items and recognize which ones are obviously "no," which ones should be a "no," and which ones you might be tempted to say "yes" to. The ones that are tempting are the ones to scrutinize and if they defocus you, if they simply do not belong because they are not part of your Market Entry Strategy, then those must be a "no." You'll probably find that it is much easier to bucket the many decisions into "no" and "yes" ahead of time before someone actually comes knocking on your door presenting an opportunity that does not fit.

A friend was the CEO of a company whose entry market was focused on breaking into mid-sized enterprises in the United States. They were early in their market entry with just a handful of customers and minimal revenues. Alibaba came knocking and offered them a $3 million contract opportunity. This offer was a pleasant surprise and came at a time when the startup could really use that money to extend their runway. Although it took some lengthy conversations, I was able to convince the CEO to politely decline. He was able to imagine the distraction of supporting AliCloud (their version of AWS) and the effort to make it work. They would have to translate the user interface to Mandarin and develop a significant support organization in China. That would be tantamount to literally tearing apart this startup's Market Entry Strategy, closing off many other avenues down the road, and would have made the company unfundable at the next round. By creating a "no" list early in their life, the startup's offer from Alibaba would have been an easy and quick decision. Several years later, the CEO still mentioned how grateful he was that they didn't pursue that opportunity.

By creating the "no" list early in the life of your startup, you'll find that when these moments do arrive, you won't have to make a spur-of-the-moment decision and there won't be any emotional arguments that can sway you. So, when your salesperson comes back from a trade show with a number of promising prospects, you'll be able to quickly respond—without second-guessing yourself, "No, we are not interested in pursuing something in Brazil. It is not part of our Market Entry Strategy." Or "No, we do not currently try to accommodate the needs of the top five US banks because we are focused on regional banks." You are trying to reduce the difficulty of your decision ahead of time, to exclude what you can exclude now before you are faced with it. The "no" list provides discipline about the critical decisions you have made and prevents you from making poor decisions on distractions outside your own Market Entry Strategy.

It is important to write down the list of "no's" that you created, and to keep it current. This is not a one-time exercise that you do as an early startup. It needs to be accessible and updated whenever you come up with another idea. It is not a guaranteed safeguard for making a bad decision, but it will be much easier to say "no" when an opportunity that would defocus you comes knocking because now you have actually considered that situation and in a clearheaded thought process, decided that you would reject it. Of course, it is possible that you could change your mind at that moment and still move forward, but you will be aware that you had previously thought about it and recognized that it would defocus the company.

EXERCISE

CREATE YOUR "NO" LIST

Make a detailed and exhaustive list of what you will say "no" to in the context of having a clear Market Entry Strategy. List the elements, which might include:

- Type / size of customers outside of your entry market
- Verticals outside of your entry market
- Geographies outside of your entry market

- Use cases that do not demonstrate your unique benefits
- Use cases where you do not have table stakes and developing them would not serve your entry market
- Supporting platforms and third parties that are not part of your entry market
- Strategic alliances / investments that are not relevant to your entry market
- Opportunistic hiring not relevant to your entry market
- Product roadmap deviations outside of your established bases of competition

THE DIVERGENT ORGANIZATION

Absent a "no" list, when some distracting opportunity presents itself and you haven't imagined or considered it, what happens? How do you project yourself into the future and try to understand whether that is a defocusing opportunity or a good opportunity? Do you nonetheless decide to pursue it or think that you can afford to take a chance on it? The Market Entry Strategy is a framework that can prevent you from pursuing something that will end up being a distraction and if you stick with this methodology, you'll minimize the temptation and the risk of making a bad decision.

So, another way to stay focused, beyond the "no" list, involves looking at your startup today and projecting what it will look like in the future if you pursue an opportunity. In this case, begin by drawing the organizational chart of your company today, using the classical representation of boxes with the reporting structure, and then sum up the headcount in each department. Do this for the entire organization, all the way to the CEO. Next, draw your organizational chart as you have imagined it will look two years later if you do not pursue this opportunity. This is based on a predictable growth model you can project because you are focused on your strategic and operating plans and not pursuing opportunities outside of your Market Entry Strategy. For reasons that will be discussed later, this plan-of-record forward-looking two-year plan is something that you should always have planned. Finally, draw the organization chart again for two years later, now assuming that you execute on the opportunity and again sum up the headcounts. Here you should recognize that regardless of the success of the new initiative, it will require headcount to be dedicated to the new opportunity. Whether it is a product extension with associated engineering costs, or an additional go-to-market direction that will result in a dedicated subgroup in marketing and sales (and perhaps a diverging product roadmap), or a new geography that will eventually require support and customization to the needs of that market—whichever it is, the key is to be fully honest about the required headcount twenty-four months into the future following your decision to pursue that avenue.

If you pursue an opportunity outside of your Market Entry Strategy, predictably one of two things will happen: either you will increase the overall headcount and burn rate, or you will divide your resources between your entry market focus area and the new opportunity. It is almost guaranteed that your organizational chart will now show a split that wasn't there had you not taken the opportunity. You will most likely end up with a *divergent organization*. It is not just the headcount split that will negatively impact your startup since you have divided your forces, but you're no longer aiming for the one beach, for the one entry market. The resulting issues and challenges you face will then come from a framework that is not cohesive and your startup will therefore not be tightly aligned to one objective. You will do *this and that*, you will have sales development representatives divided into two groups with different messages, different positioning, and different battle cards. Alternatively, you may decide to keep your sales team as one group, but then you have tasked them with the added burden of figuring out which customer fits which branch of the two directions you have created.

One of the most common defocusing opportunities we see for startups is geographical divergence. The temptation is both great and continual and there is an additional fear that if you don't address other geographies or foreign markets, some of your competitors will get a foothold there instead of you. History has proven time and again that in almost all of the

B2B markets, thoroughly winning the US market is the winning strategy since the US market is often crowning the market leader. If the US recognizes you as a market leader, you will almost always be able to conquer, or reconquer if need be, the foreign markets at a substantially lower cost than if you attempted to cover them before you thoroughly won the US. And the divergence of some resources early on will take away from winning the US, usually a key component of most B2B companies' Market Entry Strategy. The exceptions to that rule are rare and typically only apply to entry markets where US companies have no real presence.

Focused Vs. Divergent Organization

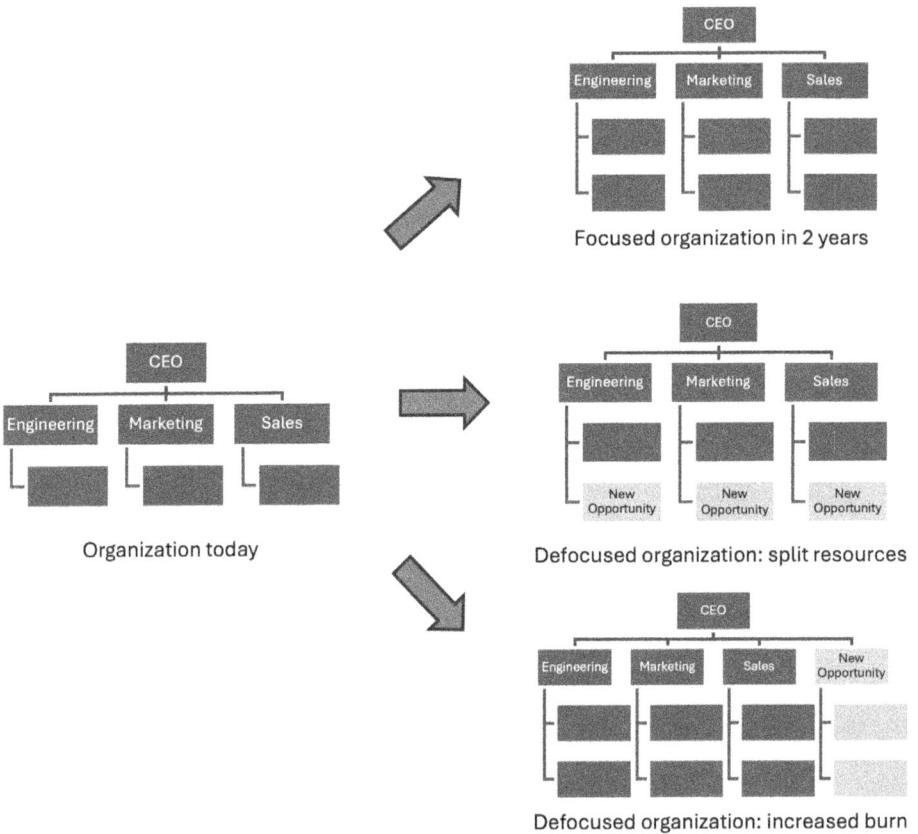

Focused organization in 2 years

Organization today

Defocused organization: split resources

Defocused organization: increased burn

I have been involved with several companies that made the common mistake of going after international opportunities too early. Someone in Brazil, Singapore, or Japan found them and said they heard about the company and wanted to buy the product. It's incredibly hard to resist that temptation for most startups and so the companies accepted the offer, given the minimal effort that seems involved. Yet undoubtedly, that customer requires support; they require local customization, different third-party systems connections, different privacy constraints, etc. Consequently, the use case represents a drift and soon after, they have a divergent organization. In one particular case, the company had gone international, yet shut down just about all their foreign efforts after they had crossed the $100 million revenue mark, simply because despite their size, it was still significantly more effective to shut that down, retreat, and deploy the freed-up resources on US-focused efforts.

If you pursue a geographic expansion too early in your growth, you will build a divergent organization, your metrics will be split between two "organizations," and you will constantly arbitrage if one part of the organization is doing better than the other. You will have to weigh which part deserves more resources. Or, as many entrepreneurs say, on which one you will "focus more." The irony of those words is probably not lost on you.

Saying "no" is one of the biggest challenges that startups face because in saying "no" you are cutting off possible unexplored avenues that might be fantastic, but you are cutting them off while you are still unsure whether the path you have chosen is the right one. It is not just startups that face this challenge, but any new product launch, any new "thing" to the world could be launched down multiple possible paths. But just as the Allied Forces did everything in their power and focused all of their resources on one shore, there is no alternative for you than landing on your Normandy beaches. The Market Entry Strategy is very much like that because you choose your entry market and the associated unique benefits on which to put the spotlight, you commit to that, you focus on that, and you do not retreat from that. By developing a "no" list and doing a thought experiment on organizational structure, you have two different guardrails to help you when the temptation to pursue something outside your Market Entry Strategy undoubtedly occurs.

EXERCISE

OPPORTUNITIES OUTSIDE YOUR MARKET ENTRY FOCUS

If you are faced with a compelling opportunity outside of your focus plan, use the following to assess the "go/no-go" decision.

1. Draw a picture of your current organization.

2. Draw the evolution of your organization in your plan of record if you say "no."

3. Draw the evolution of your organization if you say "yes."

4. Assess the spending change if you say "yes."

5. Assess the messaging impact if you say "yes."

6. Assess the sales confusion if you say "yes."

7. Assess the cost of building a divergent organization in comparison to the possible opportunity.

KEY POINTS

- Staying focused is one of the biggest challenges that early-stage startups face.

- A way to stay focused is to create ahead of time a comprehensive "no" list of all the opportunities and situations that, if they arise in the future, you will say "no" to.

- Pursuing an opportunity outside of your Market Entry Strategy will either increase headcount and burn or will diminish the resources focused on your entry market.

- Projecting the organizational impact of defocusing opportunities will help you visualize the consequences and support your decision-making.

14

Alignment by Construction

"Building a visionary company requires 1 percent vision and 99 percent alignment."

—JAMES C. COLLINS,

researcher and business management consultant

There are a number of benefits bestowed on the startup that rigorously follows the methodology of the Market Entry Strategy—beyond market leadership—and one of the most important is alignment. When a startup has full alignment of every function in the company, then all the energy of the company is focused on a set of goals that lead to one single objective. When a team is aligned, there is a clear side-effect of enabling almost effortless growth of the company. By empowering all the contributors in the organization, by removing the need for the executive team to be involved in the minutiae of daily decisions, the company moves faster in the right direction, together. If the team is aligned, they will make decisions on their own because they can inscribe their actions in the arc that they understand the company is following. The Market Entry Strategy also provides a framework for people within an organization on how to resolve nearly every question that surfaces by simply referring to the unifying goals and understanding how to react in a way that best advances the company. If you execute this approach correctly, you will be in a position to align everything that happens in your startup to all the different elements of your strategy. And, in fact, to succeed in your entry market, you must align the organization—it is not an option, it is not a "nice-to-have." It is something you must make happen to succeed in your Market Entry Strategy. Recall how we began this book recognizing that lack of execution and alignment are some of the critical startup failure mechanisms.

ALIGNMENT ON POSITIONING AND MESSAGING

Your positioning and messaging will of course reflect the mission and vision you have formulated, the entry market you have chosen, the benefits you have decided to spotlight, and the total product market requirements you have defined for yourself and for the market. If you follow the Market Entry Strategy, every element of the company and every touch point with the outside world will absolutely be in 100 percent alignment—all the time—with those key points. One commonly accepted truth is that *an excellent company executes everything with excellence,* so it is important to track relentlessly every document you issue to ensure that you are consistent in what you say in your positioning and messaging. That means tracking every collateral you print, every discussion you may have face-to-face or online, and every interaction in which you talk about your positioning and messaging. What are you tracking, exactly? The absolute consistency of the internal and external communication. Is everyone in your organization using the exact same words and key points that you have converged on? Even a slight deviation from perfect execution can lead to poor outcomes. Take hotels, a classic example in Operational Research: If every function in a hotel is executed with 99 percent excellence, pretty much every guest will have at least one bad experience during their stay because their complete experience is composed of the totality of the services they will touch. Similarly, if you execute your positioning and messaging with 99 percent fidelity, every person who sees your company's materials will see at least one deviation from your intent. So, alignment with everybody in the organization, and with every statement, is critical.

I wish I could say I was involved with only one company that was drifting on messaging and positioning, but the reality is that drifting on messaging is very common. If you review the banner of any company booth at a trade show, the benefits claimed in the demo, the messages on the website, the various posts on LinkedIn and in blogs, or the company presentation, I guarantee you in 100 percent of the cases you will find drift. Pick a company, yours or another one, and take a fine-tooth comb and see what is aligned and what is not.

Given how prone companies are to fail in strict adherence to messaging, I will add another time-tested solution: Train everyone in the company on your messaging and positioning. Everyone, including engineers, administrative staff, even the receptionist, and of course all sales, support, marketing staff, and the executive team. As an added measure of training, you can include the somewhat embarrassing practice session where each person has to repeat your messaging and positioning—at least your one-sentence—in front of an audience. This isn't punitive, and you can even make it into a game (for example, by doing a contest at an all-hands meeting). You can also prevent messaging drift by regularly listening to someone on a customer call, or doing a demo, or just describing who they work for in a social setting. You may be surprised.

EXERCISE

ALIGNMENT ON
POSITIONING AND MESSAGING

Alignment on your messaging and positioning is critical. The exact words you have chosen must be used consistently in every media.

1. Describe how every department of your organization is aligned on positioning. Do a 360-degree review of your public and private documents.

2. Describe how every aspect of your organization is aligned on messaging. Do a 360-degree review of your public and private documents.

ALIGNMENT ON TOTAL PRODUCT MARKETING

Alignment also impacts product features, and as discussed earlier, your product should have all the elements needed to demonstrate the stated benefits for your entry market. Any unclear product boundary and feature creep should be relentlessly trimmed, otherwise your startup will be evaluated on criteria where you are not the market leader. Constantly ask yourself if the features and requests by customers are actually central to your business benefits' demonstration and if not, trim them. Even if some functionality has already been developed, you need to keep asking whether any added functionality belongs in the product because if you add something to your product that is not a key differentiator, you will lengthen your evaluations and dilute your singular advantage.

I was involved with a company that had to integrate with a number of open-source operating systems. Unfortunately, as is often the case, every open-source operating system had numerous versions simultaneously present at a possible prospect's company. Rather than make a clear decision on what we would or wouldn't support, we were caught in an impossible treadmill of integrations, almost a different one for each customer. The end result was that we could never hone our product boundaries and never solidify the product. We faced an impossibly high number of different support challenges.

ALIGNMENT ON PRODUCT ROADMAP

Alignment also factors into your product roadmap, for example, when you are planning for the future (and we will see how to constantly maintain a plan in future chapters). The roadmap must be aligned to what you stated in your Market Entry Strategy. Your main effort should be to continuously increase your differentiation and strengthen your unique benefits, even at the expense of adding capabilities to the product that might increase short-term revenue opportunities. Keep in mind that your entry market customers are buying your product for the unique capabilities you have, and they may be tempted to switch to a large, existing vendor or another startup if either one eventually catches up to you. So, if your roadmap follows from your Market Entry Strategy, you ought to be able to keep your advantages vis-à-vis your competitors. You do not want your customers, faced with competing products, to decide that your competitors are "good enough."

ALIGNMENT ON CUSTOMER SELECTION

The Market Entry Strategy also informs your customer selection; the customers you choose should align with your strategic plan. This means that your marketing and sales efforts can be specifically focused on a narrower market and you can optimize your efforts and your spending to find those customers. You select the customers, not the other way around. And if by chance you happen to catch a customer outside of that entry market, you should first ask yourself why. Is there something that can be improved in your positioning and messaging such that those customers would not have been targeted inadvertently? Is there something that should be changed so that customers outside your entry market would self-deselect by reading how your focus is not on them? We have discussed what to do if customers show up who are either the wrong customer because they're not in your entry market, or the right customer at the wrong time. For now, in your Market Entry Strategy, a narrow market with differentiated benefits to select customers is what you should strive for.

A litmus test on customer selection is to play a game I call "Which One Doesn't Belong?" In this game you look at your customers' logos and see if they make sense. You might not realize why you should care but during market entry, your future investors will mentally scratch their heads and ask themselves—or you—what you were thinking when you engaged with some of them. Similarly, prospective customers will look at your web page listing your existing customers and will struggle to see if they fit your typical customer profile. I once helped hire an executive away from a seemingly successful company into a fledgling startup simply by looking at the former's customer page and pointing out to them that given the logo dispersion, their company must be totally misaligned and the internal execution confused. And I was right.

ALIGNMENT IN TECHNICAL PROOFS

By following the Market Entry Strategy you are now at the point where you can pierce the loop of "I need customers to get customers" because you now have unique benefits that specific customers want. The logical next step is to work with customers who will want to prove to themselves that you can do what you claim you do. To verify that, they often will want to execute a proof-of-concept (POC) or proof-of-value (POV) or benchmark of your product. (Note: We will expand on these terms in the following chapters.) The idea you want to enforce is that behind a POV is an objective, rigorous, and pointed experiment by the customer on your product—it is not a generic experiment to see if customers like your product and like what you do. It is not asking customers to try your system so that you can find out if one aspect or another catches their fancy. You have picked your entry market customers because they want the unique benefit that you and only you can deliver. Therefore, by construction, the customers you identified should want to test your product in their specific situation or environment. That is the essence of the POV: to prove in that specific situation that your product delivers the specific benefit, after the prospect and you have already agreed that if you can indeed prove it, they will become a customer.

While nearly every startup needs the POV, many are hesitant, reluctant even, to ask customers to test their product in a way that's aligned with those principles. You would be surprised how assertive you can be in that process if—and only if—you have rigorously applied all the other steps in the Market Entry Strategy. Importantly, don't fall into the trap of having your salespeople view the POV as the first step of the customer engagement. It has to be pretty much the last step or else everything we've carefully crafted to this point was a waste.

> **Misalignment between startups and those initial customers is common, but one instance stands out vividly. One of my companies, eager to secure more bookings, engaged with a Fortune 500 company that was only interested in the intermediate format they had after ingesting the raw input. Despite my advice to remove them from the pipeline, the company proceeded anyway. After a long and painful proof-of-value (POV) process, the customer ultimately reached the same conclusion: it wasn't what they needed. While this is an extreme example of misalignment, I'm sure you've seen plenty of less obvious ones too.**

One caution when it comes to the POV: Be ready to walk away if you discover at this stage that you have a mismatch. If the customer suddenly is no longer interested in those benefits you identified as the central point of the POV, then you'll engage in a futile exercise where you know you won't put your best foot forward. You will be tested on a capability where you know you are weak, or average, and you'll end up spending an unreasonable effort trying to win at something where you are not the market leader, where you are not better than a competitor. And if by chance you happen to win, you'll end up with a customer using your product for a use case that does not match your intent, with all the negative side-effects previously discussed.

If you execute the Market Entry Strategy correctly, you will be able to work with a customer and propose to them in great detail how a capability like yours should be tested. After all, you are the market leader and should know better than the customer how to evaluate and test your product. We will revisit this element in more detail in the next chapter.

EXERCISE

ALIGNMENT ON POC/POV

1. Describe how your definition of the POC / POV is aligned to demonstrate your unique business benefits.

2. Describe how your current customers are aligned with your use case and your unique benefits.

3. Describe how your POC / POV process is used during the correct step of the sale.

The Market Entry Strategy is not a set of disjointed ideas and street-smart tricks that improve your chance of success. On the contrary, it is a methodology that will create rock-solid alignment, the most important glue that brings all the aspects of what a company does—from engineering to product to marketing to sales—under one unifying umbrella. That operating system you create through the Market Entry Strategy is incredibly valuable to a startup by itself—even if you figure out that you made the wrong choice in defining your entry market or even, for whatever reason, your market entry ends up slipping out of your fingers or the rug gets pulled from under you. Beyond an operational blueprint, the value of the Market Entry Strategy is in creating alignment around a coherent set of directions that encompasses positioning and messaging, product features, product roadmap, and customer selection, and demonstrates your proof of value. You will now have a way to execute in harmony as well as a way to constantly measure if your startup is going in the right direction. It will be much easier to read the feedback from every part of the company if it coalesces into a single coherent framework, thereby accelerating your growth.

This is a good time in implementing the Market Entry Strategy to pause and consider whether you have 100 percent alignment. This is not the point to forge ahead, to work harder, to force things to conform to your ideas about your product, team, messaging, or positioning. It is much better to figure out if there is complete alignment on the Market Entry Strategy steps you have taken so far and if not, to return to those steps and work on them until you have alignment.

KEY POINTS

- When a team is aligned, there is almost effortless execution of the company's plan.

- Align and train everyone in your company on messaging and positioning or you will see message drift.

- Alignment also applies to total product marketing, product roadmap, customer selection, and executing POCs. Any misalignment will dilute your efforts, waste resources, and confuse your employees and the market.

- Excellent companies execute everything with excellence, so strive for 100% evidence of your alignment.

PART IV

SELLING IN YOUR MARKET ENTRY

.

15

Figuring Out Sales

"Approach each customer with the idea
of helping him or her solve a problem or achieve a goal,
not of selling a product or service."

—BRIAN TRACY, author and motivational speaker

There are a lot of misconceptions about sales. For example, people who are new to sales or have never been involved in sales often mistakenly believe that selling is a relentless effort to peddle a product until the prospect finally buys. We have all encountered the hard-charging salesperson and, regrettably, some of us have made a purchase out of their sheer pressure that we'd like to return. But the art of selling couldn't be further from that notion of convincing someone to buy. Sales is really about these four points:

1. Finding people who have the problem you can solve

2. Understanding thoroughly the benefit they would enjoy if that problem were solved

3. Demonstrating how your product is uniquely suited to address their problem

4. Helping fit your product into their processes

Your goal in sales is to help customers go forward because of what you have enabled them to accomplish. These four points are directly related to the Market Entry Strategy. Selling is first and foremost being skillful as a business advisor to your customer. Those people who fully embrace these keys of understanding your customer and being an advisor to them can sell again to the same customers with whom they have developed a bond of confidence. Not surprisingly, most sales training programs reinforce the notion that the most important skill of an effective salesperson is to listen most of the time and to seldom speak.

It's good advice. With technology advancements widely available today, there are tools that allow you to listen in on sales calls and then be provided with advanced analytics on the interaction between a salesperson and a customer. It's often a surprise for all when the speaking / listening statistics are analyzed because they usually show the opposite of the art of selling: more talking by the salesperson and far less listening. While there are many books specifically dedicated to the topic of sales, the purpose of addressing sales here is to examine how some of the key components of the sales process fit within the Market Entry Strategy and the core principles of you choosing your own entry market with your unique benefits and also the basis of competition at which you excel.

HOW TO SELL

The idea of a section with this title may seem trivial or fundamentally basic, but most startup teams do not think seriously about how they will sell their product until far into their startup journey, and that can lead to problems. It is also quite common that many founders have backgrounds in the hard sciences like engineering, computer science, or medical fields and their belief about sales is that the role is to push people to buy a product. Beyond having an incorrect understanding of sales, many startup teams make the mistake of selling in whatever way comes to mind with the hope that over time they'll learn which techniques are most effective. The idea of randomly experimenting around sales is similar to finding product-market fit by throwing a minimum viable product (MVP) in the wild and seeing who likes it. It might work, but chances are that you won't converge to the right sales model quickly and may squander any customer progress you could have made along the way. Instead, the Market Entry Strategy advocates that before you start selling, take a moment to reflect on who are the customers needing your product the most. How do you find those people? How will you understand and demonstrate the business benefits that would result from them adopting your product? And how will you help customers fit your product into their workflow? Those are the critical questions to answer *before you approach any customer.*

> **I was involved with a company that applied this approach from day one. They actually had a clear understanding of who the customer would be, what business benefit they would demonstrate, and what the sales process would be before they wrote a single line of code. This clarity shaped their total product marketing to align with their sales approach. They relentlessly debated the product boundaries that, if not enforced, would derail their envisioned sales methodology. The result was one of the most successful market entries I've witnessed.**

When it comes to finding potential customers, several approaches are available. For example, one approach is to attend trade shows where you think your customers will be and try to meet them there. Or, if a trade show isn't in your budget, you can generate original content and publish it in relevant industry magazines. If you have intimate knowledge of the key opinion leaders in your industry, you can try to pitch to them, or hire people with access to the decision-makers you'd like to meet. Whichever approach you think will work for your startup, it should be grounded in a thought process that led you to that conclusion. It is not trial and error, it is not chance, and it is not whoever your personal network or your investors happen to be able to access. You need to examine what has or hasn't worked before in your entry market. And, knowing the entry market you chose, knowing your vision, mission, strategic, and operating plan, you should be able to design an approach that will put you in front of your target customers. If you only have a hypothesis, if you are not sure how to find customers, then at least proceed in a structured and organized way rather than in a haphazard fashion. If you are methodical in solving the "finding potential customers" problem, you'll be able to better test whether your approach is effective or not.

Depending on your product, on one side of the spectrum, you might have an easy way to insert yourself into the customer workflow, where the only thing a customer has to do is press a button to have your product seamlessly activated. But on the other end of the spectrum, you might need a lengthy process of obtaining permissions, connecting databases, and integrating with other products. If so, your product might disrupt or intrude on the customer's current way of working. Before trying to move to selling mode, it's best to analyze the impact of your product on the customer's workflow so that you know what you're dealing with. You cannot be successful with a sales campaign if you do not have clear answers to those questions that will inevitably be asked by your customers early in the process and will test your credibility as a new supplier.

Another issue to think through before actually selling to a customer is what it will take to demonstrate your unique benefits—not your product features or functionality, but your customer's *business benefits*. All the technical breakthroughs in your product—the artificial intelligence, big data analytics, finely honed heuristics, or proven-optimal search algorithms—all of those do not really matter to the customer unless your product solves their problem. If some of the cool features in your product do matter to the customer, it is most likely only in the context of having differentiated results when it comes to business benefits. So, the art of sales is not about you and your product, but about you relentlessly putting yourself into the customer's shoes and asking, "What is the business benefit?" Whatever those benefits are, they need to be thoroughly thought through and documented even though you may not have a high degree of certainty around them, especially if you are serving an emerging market segment or solving a recent problem. A working idea or model of how you will find, understand, and provide benefits to customers is critical to the next steps of figuring out sales.

WHOM DO YOU SELL TO?

Up to this point, the term "customer" has been used in a general way but it is important to understand that *there really is never one customer*. For every product, and for every organization you identify as a customer, there's a specific set of people who are the buyers—and they are not homogenous, they do not have the same interests, and they may not be in the same department. Few products have a single decision-maker where a person can decide on their own whether to purchase a product, take out a credit card, and make a purchase. Even in the rare case where there is a single decision-maker, the elements identified above, like finding customers, demonstrating the benefits, and understanding how your product fits into their workflow, still apply. For the vast majority of products, the customer resides somewhere inside an organization and you'll need to navigate and understand that organization so you can find all the people your product impacts, positively and negatively.

An example of a list (and by no means exhaustive) of people who could be involved in a decision to purchase your product include:

- The **specifier** is the person who decides what specific functionality they're seeking and what key indicators the product must perform against. The specifier is most often someone with a technical background, or in a technical part of the organization, but not always, so it's important to not assume that they're the first technical contact point.

- The **user** is the person who'll eventually be the one using the product. The user can be the same as the specifier but in many cases, these are two different personas. Startups tend to focus on the user, believing if they can convince the user, they will ultimately get the sale, but the user is just one of several people impacting the decision about whether to purchase your product.

- The **budget holder** is the person who eventually signs off on the purchase. They have the ultimate say about whether the product gets acquired since they're typically accountable for the return on the investment they've spent with you.

- The **decision-maker** is sometimes the user, sometimes the specifier, but oftentimes the decision-maker is not directly visible to you. This person may be a high-level executive who reviews presented alternative solutions or reports justifying the purchase. Note that in certain cases they are different from the budget holder.

- The **purchaser** is the person issuing the purchase order. Sometimes they're working on behalf of the decision-maker, but in large organizations the purchaser has the power to negotiate or even block a purchase.

- The **blocker** is someone who can negate all your efforts. This might be a compliance person, or someone with a technical reputation you haven't been able to convince, or sometimes they are simply someone working for a parallel organization that can persuade others to standardize on a different product than the one you're proposing. Assume there's always at least one blocker and even when you've found that person, keep looking for others.

There might be variations on the roles identified above but it's best to assume that the main job of a salesperson is to *identify the people in these roles within an organization*, listen to them, understand their key motivations and objections, and figure out if your product will be recognized as solving a key problem. Even then, the salesperson will need to understand how to demonstrate the business benefit and how to fit your product into the customer's workflow.

I was once involved in an organization that was developing an important strategic relationship in which my team
was delivering a key product. In every presentation we
made, the technical team who would be the user, the decision-maker, and everyone to the most senior executive
level, all were nodding their heads in agreement with our
proposal. But there was one person in the back of the
room who would stay quiet until the very end of every
presentation, then throw some highly negative technical comment about our offering. We never found out
this person's name, we didn't know what their role was,
or even what department they were in. We nicknamed
this person "the crazy PhD in the back" because we as-
sumed they had a PhD, but we never managed to sell our
product. Beware of the blocker, even if you have wide
agreement from everyone else in the organization you
are selling to.

For example, for most products, you will have a technical decision-maker who may or may not be the actual user. If that is the case,
you might have to convince the people who specify what gets bought, but
unless the end users accept your product, it will not stick. If you make the
sale anyway, the disappointing usage will still cause the specifier to not
renew the contract—why would they? Nobody is using your product.
There are countless examples of startups getting tripped up because of
a disconnect between the specifier and the end users. For example, you
sell a security product that a customer needs—and meets their specifica-
tions, but your product is introducing friction into the workflow, so the
end users stop performing certain tasks or find ways to circumvent the
system. Or, your product provides benefits to Human Resources (HR),
but there is minimal adoption by the employees. More recently, software
developers have proven to be a particularly stubborn group for product

adoption and we have seen that a majority of products bought by others on behalf of developers to boost their productivity, reduce bugs, or make their code more secure are simply never used.

> Most startups have some form of weekly sales pipeline review. The #1 goal of this review is to interrogate the salesperson but also the other stakeholders to find out if they have identified the people in these roles and can explicitly articulate their business benefits and objections and plan the actions they will take to fill in whatever is not yet fully understood. There is no deal that has any certainty of closing without that mapping of the account and when a deal falls through, you can almost always trace it back to such a gap. Keep in mind that it's natural for salespeople, as human beings, to be afraid to ask the questions that would invalidate all their efforts so far. One way to overcome this situation is to make sure to celebrate someone who has disqualified a deal in the pipeline because they mapped all these people and discovered that there was no deal to be had anyway.

As part of your discovery process, remember that you're looking for the actual business benefit, and the actual expression of that benefit may be something that the people you believe are the decision-makers do not see or do not care about. In this situation your best option is to identify the person within the organization who cares about the business benefits you provide and make that person central to your sales effort. For example, if you have a product that reduces overall downtime, that may be a business benefit to someone who cares about being able to execute a task they're responsible for without interruption, but it might not be perceived as a benefit to the person in charge of working on a specific system and for whom you just created more work. Or suppose you have a product that indirectly improves a specific collection flow but the financial benefit to the enterprise is much more substantial than what the purchase specifier recognizes within their department. In both cases, if you are not selling to the person who sees and fully understands your business benefit, you aren't going to make as much headway. So, you need to find the person in the organization who cares the most about the business benefit and include them in the sales process.

After you find the person who will benefit most from your capabilities, you need to understand who can object to what you are offering; who can derail, delay, or outright reject your offering? Oftentimes a blocker can be someone you've never met and who seemingly has no understanding of your product. For example, any large corporation has a compliance department with veto power over your product—no questions asked and no recourse for you. Large companies also have a vendor onboarding process that often is onerous, vexing, and a hurdle for the startup that's difficult to overcome. Some companies may also have internal processes that involve issuing a *purchase request,* in which the internal purchase decision needs to be justified with the same completeness as you'd need to get clearance on a purchase order.

One of my very successful companies was trying to sell their product to a large telecommunications company. Several other telecommunications companies had already bought the product without any compliance objections. However, this particular prospect recognized that in a corner case, someone with the appropriate access privileges could briefly and indirectly see their end customer data unencrypted. It didn't matter how many times this was explained away and despite the fact that everyone in their industry had already become comfortable with and was using the product, we could never overcome that objection.

Another challenge startups face in selling to large organizations is privacy, and if you do not fully understand or comprehend the direct or indirect legal exposure of privacy issues, you will not have your product adopted. For example, imagine you have a system that uses consumer voice interaction with your customer to optimize some process. At face value that seems like a really good product, but most companies will not want you, nor their company, to have those voice recordings stored—not on their side and not on yours. The person you think is making the purchase may not even know that such a privacy concern exists and yet it can still prevent the sale.

Another common hurdle for startups to overcome is a rule in many companies that prevents overreliance on a single supplier. In some cases, a company will dual-source products or restrict a product to a certain penetration level while ensuring that another solution is used for the rest of the corporation. And some companies are just allergic to other products upon which you rely as part of your product, maybe because of their country of origin, or their legal status, or simply because they have a competitive relationship with that company. The list of issues that can thwart your successful sales effort is long, and one big part of being successful is the ability to understand where the people who can block you are, and what their issues are.

One of my companies had gone on a multiyear campaign to sell to one of the largest banks in the United States. We overcame all the technical hurdles, the balance sheet and company viability scrutiny, the web of integrations they required, the security-compliance concerns, and the list went on. Finally, the customer issued a very large purchase order and champagne corks were ready to pop. Then one of the cloud providers upon which our product was built made a move that rendered them persona non grata for this customer. They told us they wouldn't withdraw the order but would not pay the invoice until the product was available on another cloud.

Finally, if you have a product that requires a specific configuration, one that is a large-ticket item, or one that requires a profound transformation of your customer process, you may require sponsorship of a high-level executive with the authority to make those decisions. What characterizes situations where you need high-level support (and maybe more often than you think) is the personal motivation of the person whose sponsorship you seek. In nearly every case, you'll have to discover what motivates that person and how your product capabilities enable them to demonstrate to the upper management, sometimes to the CEO or their peers, that they have exceptional leadership. It is part of your process to understand who that person is and what business outcome your benefit enables. In those cases, having a *vision match* between your startup upper management and the potential customer's management team is a critical part of the sales process. If you are aligned on how your solution will transform their business, it will dramatically impact the opportunity. If you invest in gaining that alignment, it will have more influence on whether or not your product is adopted than just about anything else you can do. Of course, recognize that at the C-suite level, an incremental difference does not cut it and if your product falls in that category, you'll struggle to get the attention of upper management. Sometimes you might have to change your product to enable that benefit to be presentable to the executive level—it could take the form of a specific report or a trend line that demonstrates the before and after, but think through what that person will want to present. A variation on that theme is an executive who's financially compensated to achieve a specific management objective. You would be surprised how small an incentive it takes to get a person to get behind your product if it helps them reach that goal.

An important word of caution about relying on an internal *champion*. If you've effectively identified and convinced a specific individual that your solution is right for them—along whatever axis that person cares about—they may champion you and your product and walk your startup through the different internal steps needed to complete the sale. At first glance, having a champion may seem like a great thing for a startup and indeed, many business books advocate identifying a champion. But working through a champion is a lazy substitute for the Market Entry

Strategy sales process described here. Why? By leaning on a champion you do not actually know who the business benefit owners are, who the budget owner might be, or who could object to the adoption of your product. You may not even know if your champion is well-regarded in their company or if they are known for taking on adventurous experiments that lead to dead ends. Maybe your champion is a great person and means well, but they're also too junior or too recent in their company to fully grasp its decision-making process. By focusing on a champion to the exclusion of the specifier, budget holder, user, and blocker, your sales process has a single point of failure; this dependency on one person also prevents you from creating a repeatable sales process. If the person leaves or is reassigned, your whole sales construction collapses.

> **As a CEO or executive in a company, it's critical that you dedicate a significant amount of your time to building relationships with others at your level and use your title to justify access to the highest level of management. In the case of a prospective customer executive, they have to be able to trust that you, as a person, can lead your company to be a partner to their strategic ambitions and that you see the world via the same lens. There can be an unfathomable impact of this conversation—which is not a sales opportunity and doesn't go into your product capabilities—which focuses on your vision and mission, and which is also validated by your strategic plan. The executive with whom you will have a vision match and who recognizes you as market leader and business advisor is a tremendous asset in winning this customer and others in their industry.**

One last caution about aligning your startup's fate with a champion is something people understand only when it happens to them: you *owe* your champion. Your champion can make demands, such as adding new features, and you'll be forced to accommodate them—even if they don't align with your roadmap or aren't critical to this customer—because you cannot afford to frustrate the only person in your favor. These demands can take many different forms and can originate from well-meaning intentions, but the situation always eventually backfires. So, while the idea of a champion is appealing, do not fall for this lazy substitute to a carefully orchestrated and comprehensive sales process.

EXERCISE

WHOM DO YOU SELL TO?

1. Identify people with the following roles within each prospective customer of your entry market:

   - Specifier
   - User
   - Budget holder
   - Decision-maker
   - Purchaser
   - Blocker

2. What is the motivation for each person to work with you?

3. What are the benefits for each prospective person working with you?

4. What are the reservations each person has about working with you?

5. If you can't identify one of these stakeholders, what action are you taking to find them?

6. If you are currently relying on a "champion," how are you lessening your dependence on them?

WHAT IS A PROOF-OF-VALUE, AND DOES IT MATTER?

Over the past several decades as software and especially software as a service (SaaS) have become more prevalent, it has also become a common practice by startups to include either a proof-of-concept (POC) or a proof-of-value (POV) analysis in the sales process. I intentionally use these terms interchangeably, but most people see the POC as a light exercise early in the sales cycle and the POV as a deeper exercise later in the sales cycle. The key point here is equally valid: be careful not to engage your precious technical resources until you've made sufficient progress in your sales process. Note that while these value demonstration exercises are common, they are too often seen as the end-all or the be-all for all products. Many products can be delivered against a specification or a set of known metrics that the product has proven or committed to deliver. Some software products can be sold "as is" and with a contract that typically includes an initial period, paid or not, that allows the customer to terminate the agreement. Sometimes customers can cancel the contract for convenience and other times—the most common reason—because the product does not meet its claimed merits. Before you blindly follow the route of a POV, consider if those alternate routes might work for your specific product and environments.

However, if you decide to include a POV in your process, it's critical to recognize that a POV is not performed early in the sales process and shouldn't be a (lazy) way to entice the customer to engage with you. Instead, it is best to follow and complete all the steps of the sales process and leave the POV to almost the last event in convincing the customer. Why? Because if you can't first convince the customer that they're buying a set of business benefits they care about, that they're getting a solution to their problem that needs to be included into their workflow, then no POV or POC is going to convince them. You can convince a customer by listening to them, understanding their problem, and by describing your solution and benefits. If you have documented all of these steps using the Market Entry Strategy process, you should be ready first to present a compelling proposal to the customer. It's important to remember that, using the methodology, you chose the customers as part of your entry market precisely because they're the ones with the problem you're solving. And you chose these customers specifically because your product uniquely provides the benefits they seek.

You can use the Market Entry Strategy to your advantage in moving through the steps of the sale because your knowledge and your customer intimacy should prevail. The reason for a POV, if you decide to include one, is only the ability to verify in that specific customer's environment that the claims you made are indeed true and that the business benefits are obtained. The POV step should really take place only after you have found the decision-makers and naysayers, only after you've determined that a budget exists for your product, and only when you reach the point that proving the truth of your words is the only element that stands between where you are in the process and getting that coveted purchase order.

If you are providing a POV, the definition of its scope must follow the Market Entry Strategy methodology. That is, your POV should include why you selected this potential customer and that you selected them because of their need for your unique benefits. You are now in a favored position to maintain the POV focused on these elements and these elements only—and this is critical because you only want to be evaluated on the criteria you have selected, where you excel. Recall in the competitive spider chart you created that you chose the bases of competition where you dominate and where you believe your customers will care about your differentiation. The POV needs to only demonstrate that those unique features are present and that you outshine any alternative competitor trying to provide them. On the other bases of competition, you are only striving for a passing grade, having the necessary table stakes capability but no more. With the customer, devise the plan that will enable them to walk away reassured that what you claim is upheld in their specific situation and environment. Since you are the market leader in your entry market, this makes you the knowledgeable expert and the customer very often will readily accept your POV—with only minor changes or additions—on how your system should be tested. If you meet with a customer and hand them your blueprint on how to test your product, you may be surprised how often the customer will take it as an inspiration for their plans or even just adopt it unchanged.

What if the customer wants to test many other functionalities? Here, you need to be very cautious. If the customer wants to test elements of your product that are not your chosen bases of competition you will most likely fail because, by definition, they're testing you on things where you are not the market leader. Perhaps you incorrectly included this prospect as part of your entry market when it turns out that their concerns or interests lie elsewhere. Maybe you determined that their use case fits what you had defined but now realize that the customer is thinking of a different problem to address. Or maybe the customer simply has a much broader need than the one you specifically focus on. No matter what the disconnect is between you and the customer, this is a critical time to sit down and walk again through the steps of the sales process that led you to this point.

It might still be possible for you to modify the customer's POV plan, but sometimes your best course of action is to withdraw because it would be foolish to engage with a customer where you are not evaluated on your strengths. If a customer does not care enough about your unique benefits to test you on those and instead insists that you will be tested on functionality you know you're inferior at, you will have regrets in the future. You will regret entering into a vortex that will monopolize your sales and engineering resources to make up for your shortfalls; you will also regret putting your engineering team into a frenzy trying to develop the missing capabilities that do not fit your roadmap. And even if you win that customer, you'll most likely end up with a side use case that you can't use as a reference in your marketing programs and the customer will always be at risk of churning.

Despite all the problems associated with the early use of a POV, many startups use the POV as part of the customer-discovery process and eventually find themselves in trouble. Initially, getting a POV "sale" sounds great, but it can be taken as a substitute for actual progress in your sales. Your conversion from successful POV to a purchase will be abysmal, since all the other aspects of the Market Entry Strategy sales process weren't addressed. Salespeople early in their career will often be tempted to pursue a POV agreement because it appears to signal progress, which secures their job and makes them feel valued. Unfortunately, the POV is not an alternative for the sales steps and most startups figure this out later when the conversion ratio collapses after having spent their most precious resources in vain. Hold the bar high, demand your salespeople ask all the questions in the sales methodology we described, and recognize when you're facing an unsuitable prospect rather than advancing toward an unnecessary POV.

> **The advent of Salesforce and other customer relationship management (CRM) platforms has been mostly beneficial as a repository of the customer information and the sales pipeline. However, I've also seen that those platforms come with an associated demand that every person involved in the customer interactions meticulously and continuously update their records. The unfortunate side-effect is a whole generation of salespeople who mistakenly confuse the art of selling with the advancement of their pipeline through the CRM records. Make sure your organization prioritizes the actual selling motion over being "CRM jockeys" measured on abstract metrics.**

HOW DO YOU SUPPORT THE CUSTOMER?

Most startups are often entirely focused on the pre-sales motion and are putting all their efforts into the steps specified above. For many startups, support is almost always an afterthought, considered after the purchase has been completed. However, waiting until after the purchase of your product is a perilous approach. It's perilous because you may find that your product is structured in a way that the support—and your path to success—is extremely complex. This complexity often stems from things you haven't considered, like assuming you'll have certain types of access or permissions but those are things the customer never intended to provide to you.

> A startup I was involved with faced a tough competitive situation against another startup that had built a much more simplistic, less elegant, and poorer performing system than ours. When our product was running without a glitch, it was much easier to use and delivered superior results. But our system was monolithic and theirs was made of small components where skilled application engineers could tune the results, sometimes to the point of manually forcing their system to produce the output their algorithm was incapable of finding by itself. It didn't matter that I could make presentations that demonstrated the superiority of our approach; in actual direct comparisons, our competitor always managed to squeeze out a competitive output from their system and create significant headaches because they'd built a system that was easier to support than ours.

Another element that sometimes a startup will recognize too late in the sales process is that their product's possible downtime would have enormous consequences, or the fallout of an erroneous result is so high that the customer has demands a startup will struggle to meet. Those customer demands may come with penalties, liabilities, or insurance requirements that you haven't anticipated and haven't begun to consider until they appear in the fine print of the customer contractual agreements. If you find yourself in this situation after spending considerable resources getting to this point, it might be difficult to walk away. Think carefully about continuing; some of these fine-print stipulations may effectively negate your ability to recognize the revenue. More importantly, perhaps, a subsequent investor or a potential acquirer will conduct due diligence on all of your contracts and might find them untenable. These contracts may be showstoppers for investors or acquirers whose view of the exposure of accepting these fine prints will very likely differ from yours.

The human element of providing support to your customer is of course also relevant and needs to be considered. Do you have the staff to ensure the success of the customer and the delivery of the benefits you so adamantly touted? It is entirely possible that the users of the product after the purchase are not the same as the people who tested it, and the users will not have interacted with your team as deeply as those executing the POV. Similarly, it is often the case that your top technical application engineers are assigned to pre-sales and the wizardry they displayed may not be available to the customer after the purchase. The customer may also have demands that are not explicit (or even known by them) until they fully deploy the product. For example, customers may insist that you provide support across many geographies or require a 24×7 service level agreement (SLA) that you only hear about late in the game. Or perhaps the connectivity and permissions you enjoyed during the POV are not comparable to what users experience in their daily workflow. Whatever it may be, it's highly recommended that you make the support plan an integral part of your product and specify what that support entails early in your sales process.

As in the situation with your POV, you'll be surprised to find that many customers will be open to having you present a support plan associated with your product. Presenting the support plan as part of your pre-sales effort is a powerful enticement that alleviates customer concerns in adopting your product. Recall that you are asking the customer to change their habits and to take a risk to adopt your product, and there's often an unspoken fear that will not be brought up in your conversations, yet it is very real. You gain a substantial advantage by having a coherent support plan and sharing it as early as possible. Even sharing your typical support plan during your very first meeting with the customer will make a big impact because they'll understand that you are able to make support an integral part of how you'll deliver your unique benefits after they purchase your product. That goes a long way in reinforcing your startup as the market leader, and providing the customer with assurances that you stand by your product and any issues that may emerge after it is deployed.

EXERCISE

ELABORATING YOUR SUPPORT MODEL

1. Identify all the integration elements for your product to be deployed with a customer and used in production:

   - Permissions

   - Access

   - Integrations

   - Privacy

   - Data

   - Legal issues

2. Identify all the support elements for your product:

   - Post-sales customization opportunities

   - Support processes not requiring engineering involvement

   - Maturity of the product in similar production environments

   - Ability for a junior support engineer to intervene in most situations

   - Detailed escalation process

3. Write an example of your support plan that can be shared with prospects as part of your pre-sales.

PRODUCT LIMITATIONS

There is a lot of advice for startups that they should "fake it till you make it," and this has become an essential part of startup folklore. Although some people believe it's the hallmark of entrepreneurship, it often borders on the unethical and if not, it is definitely unsavory. The main argument against following that approach is that, in reality, it simply does not work. Customers will always eventually find out that your products have limitations, that your product does not live up to your hype, and then your fledgling reputation will be badly damaged. A damaged reputation is difficult to overcome, especially for a startup, and often permanently prevents you from achieving market leadership.

Besides a tarnished reputation to the outside world, there are internal challenges if you decide to fake it till you make it. Your ability to motivate an organization to behave in this way is very limited because most people are uncomfortable participating in a deception. Beyond that, it impacts the culture you advocate and the perception of your own leadership. Maintaining the appearance of a certain ability yet dealing with an actuality that is very different is thoroughly exhausting for everybody in the organization. At every level of the organization, people have to remember what was said to the customer versus the reality, and it requires a considerable amount of mental resources to make it work. Additionally, actual resources are wasted throughout the organization as people work on Band-Aids and temporary solutions.

The Market Entry Strategy advocates that your goal is to achieve early-market leadership and once you do that, to use that position as a foundation to reduce your cost of sales. You will have selected your customer because of their interest in your capabilities, and your entry market customers are conditioned (if you've done your job well so far) to focus on what you do best. You do not have to fake it till you make it because you can be honest with your customers, openly communicating with them about your limitations about other elements. You can also be forthright with them because they should care more specifically about your differentiated capabilities. If that spooks them or raises red flags for them, then you have to ask yourself: Did I properly qualify them? Are these the right entry market customers?

As a startup, you will most likely have some limitations. Either your product is not fully fleshed out, or you have never tested it at the scale needed, or you've never deployed it in the specific environment the customer envisions. You must be completely honest with the customer and work side by side with them to accommodate an insertion into their workflow that takes into account your limitation. If your product currently supports one platform they need and not another, sit down with them and work on a phased deployment plan. If you have never dealt with the scale of their full deployment—you've tested your product with thirty users, for example, and they want to use it with one thousand— you need to level with them and jointly design a progressive ramp in the number of users. If you have this open attitude and communicate with them as partners, they'll work with you and very importantly, they will respect you for your integrity.

In summary, implementing the Market Entry Strategy and having a comprehensive approach to your sales process that centers on your entry market and your market leadership allows you to guide your path to figuring out sales. Having a cohesive plan centered on understanding the customer and delivering your unique benefit to those who really need it, in an open and thoughtful approach, will make it easier for you to navigate this critical step of your market entry stage.

EXERCISE

PRODUCT LIMITATIONS

Create a list of product limitations:

1. What scale has it been tested at?

2. What scale has it been deployed at?

3. What types of environments, including third-party integration, do you not support?

4. Are there any other limitations that apply to your type of product?

KEY POINTS

- The Market Entry Strategy advocates that it's a mistake to "wing it" in sales or to experiment; instead, think carefully about which customers most need your product.

- A large number of people could be involved in deciding to purchase your product, including the specifier, the user, the budget-holder, the decision-maker, the purchaser, and the blocker. You must discover and map what each one cares about.

- Beware of relying purely on a "champion" to sell on your behalf. You can develop blind spots and an over-dependence on a single person.

- Perform a POV only when the customer has been convinced to buy and you only need to prove that you can deliver the promised benefits.

- Do not perform a POV early in the sales process as a lazy way to entice a customer to engage with you. Be ready to walk away if there is a mismatch in the POV definition.

- Recognize the importance of customer support and consider including it early in your sales motion.

- Nearly every startup product will have limitations; be honest with your customers and work within your limitations.

16

The Sales Spectrum

"Whatever you are, be a good one."

—ABRAHAM LINCOLN, sixteenth president of the United States

If you follow the Market Entry Strategy to figure out sales, you will have identified the key customers who have a critical problem you can solve and will know what you need to do to complete a sale. That is step one. You now have the foundation for building a sales team and eventually, building an entire sales organization. The initial engagements for most startups are typically handled by the founding team or by the initial executive team, and while that's convenient (or a necessity in most cases), it clearly does not scale. Where do you go from there? How do you build a sales team? One of the first things you can do is recognize that there are many types of sales processes and many types of salespeople in a continuum ranging from straightforward to very complex. On the straightforward end of the continuum, there are junior people following a script on a phone call and on the complex end, there are seasoned sales executives who will be able to map an entire account, build relationships throughout a large organization, and conclude a multimillion-dollar contract. In this chapter, we'll analyze the different types of salespeople and how they connect to the type of product you have and the sales steps you hope to execute.

THE TYPES OF SALESPEOPLE

It's important to understand that with sales, a "one size fits all" approach doesn't work that well because salespeople have different specialties, different approaches, and will thrive in the sales process that best matches their character and sales training. Although people will evolve and grow in their career and can obviously develop new skills or be trained to cross into another specialty, it's a recipe for disaster if a startup has to immediately train people with one skillset to learn a completely different one. The startup doesn't have the maturity or bandwidth to spend resources on this task. It is a mistake we see all too often in fundraising presentations: the startup team assumes they can hire any type of salesperson who will figure out the appropriate sales model and adapt to its needs. This is incorrect; it is vitally important to hire the salesperson who matches the sales process you need them to carry out and to resist hiring any salesperson simply because you desperately need one. Below are high-level descriptions of the types of salespeople you're likely to consider, starting from the simplest to the most sophisticated.

- **No salesperson needed.** It has become part of the startup ethos to consider product-led growth (PLG) as an attractive alternative to building a sales organization. What could be easier, right? The customer goes to your website and selects a product, clicks a button, and completes the sale. As simple as it sounds, the bar to clear is that you must have a product that customers can discover themselves; they must be able to understand the benefits themselves, they must be able to decide on their own that you have a solution to their problem, and they must be able to demonstrate to themselves that your product fits into their workflow. Only if all of those criteria are met, then that customer will buy your product—unassisted by your startup—typically by simply entering their credit card information on your website.

- **Remote scripted salesperson.** The next level on the continuum of salespeople is the remote scripted person, which you can think of as a junior or entry-level position (but some people find scripted sales to be their calling and spend their entire career in this role). As the term suggests, the person in this role follows a script. They could be cold calling a customer or upselling a customer from a free to a paid account or upselling to added capacity or functionality. The key point is that a person in this role needs to be handed a script or a battle card and they will have limited ability to work outside of that script (to handle objections, for example) because they are typically not technical at all. The scripted salesperson functions best in situations where a single phone call is all that's needed to execute the sale. They are almost always remote and driven by simple activity metrics.

- **Simple-process salesperson.** The simple-process salesperson is someone who tends to be early in their career (but not necessarily). This person can handle a multistep sales process, discover the customer profile, understand the needs and requirements of the customer, and set up meetings and subsequently close a sale. Unlike the scripted salesperson, the simple-process salesperson can carry out a degree of competitive analysis and can deal with customer objections. Note that people proficient in this role need to be heavily trained, have a clear process to follow, and have battle cards for most situations. Sometimes a person in this role is only used for the discovery and setting up meetings before passing the customer onward for a more complex sales process. People performing this function are referred to either as a sales development representative (SDR) or a business development representative (BDR), depending on whether they receive inbound leads (SDR) or prospect outbound leads (BDR). This role is almost always remote—though there are benefits to being part of a team that can share experiences. The simple-process salesperson is evaluated on metrics driven by measurement of activity, progress, and conversion, all which are heavily embedded into their process.

- **Transaction-focused salesperson.** The transaction-focused salesperson title often gets maligned. In our context it simply refers to a person who handles a relatively short sales process, but it could still be (and often is) a process that is complex—it just does not involve building lengthy personal interactions with customers. A transaction-focused salesperson will often manage a more complex sales process, involving multiple steps and organizational discovery. A person in this role is able to track down the different types of people involved in a customer organization (specifier, budget holder, decision maker, user, etc.) and will be able to understand if a budget exists, manage the ROI justification for your product, and help push it through legal and purchasing departments. A transaction-focused person is often provided qualified leads generated by marketing or SDR/BDR and their job is to manage the process from that point to a close. People in this role are also able to construct a multiyear contract or one that involves a progressive deployment; they can also coordinate the startup's internal resources to best serve the customer. A transaction-focused person will need clear playbooks and positioning as well as periodic training and assistance from their management. People in this role are usually hybrid, spending a significant amount of their time on the phone but also meeting customers at trade shows or on-site over the course of the sales process. They have full autonomy over how they manage their time and activity and are measured through the progress of their pipeline and the actual closing of deals and quota attainment.

- **Sales executive.** A sales executive is a high-level person who can represent the company in C-level meetings with the customer and typically have the gravitas and the depth of knowledge and experience implied by the "executive" title. A sales executive will know and be known by others in the industry they are selling to, and often have a history and familiarity with the specific customers they're targeting. Although a person in this role can take leads generated by others and close sales, they typically will also be able to build most of their own pipelines. Beyond creating a pipeline, a person in this role will be able to map the entire account organizational structure, discover the various stakeholders within the customer organization, and will be able to either personally build a relationship with the sponsoring executive or enable the startup's management team to meet them and create the vision match: the critical alignment around a common market vision and the company's mission. A sales executive should be able to build an entire account business plan that integrates the customer organization, the benefits expected, and how the company will engage with the customer —and present that plan to the startup senior management. A sales executive will thrive in situations with complex processes, contractual negotiations, and customized engagements. They have no real upper limit on the size of deal they're looking to construct and are not afraid to come up with a justification for a multimillion-dollar contract. Like anybody in a sales role, the sales executive will need clear positioning and training but unlike other sales roles, this person will also need access to senior management on a regular basis to support the vision and align their activity with the company's strategic directions.

It is important to recognize that each type of salesperson has a built-in range on the size of deal they will be able to close. I was once interrogating a transaction salesperson about the potential of a very large account they were starting to work on. To my surprise, the person thought the deal value would be under $100,000. I challenged the salesperson to imagine finding themselves in the proverbial elevator ride with the customer executive responsible for the related business. We methodically discussed the business benefit of the product on the customer revenue and agreed that it was comfortably over $100 million. When I asked the salesperson why we wouldn't be able to justify a $2 million order for the benefit we would deliver, they blushed and said, "I could never ask for such a number."

The type of salesperson you need is dependent on your product and your customer; the type of salesperson you choose requires careful thought. The sales role is not a single, fixed job description but varies along a continuous spectrum from relatively straightforward to complex. This is obviously a simplified view to help you situate the different salespeople you're likely to encounter as you build your sales organization.

KEY POINTS

- Most founding teams handle the first sales engagements themselves, but this doesn't scale.

- The type of salesperson you need is dependent on your product and your customer.

- Rather than hire any salesperson, think about hiring the salesperson who matches your product, customer, and deal size:
    - No salesperson needed
    - Remote scripted
    - Simple process
    - Transaction-focused
    - Sales executive

17

The Magic Triangle

*"Life is all about finding the
right angles and connections, just like a triangle."*

—ANONYMOUS

The Market Entry Strategy is a methodology that can be used from the very inception of an idea or product, all the way through to building an organization and scaling it. All the work you have done so far on your unique benefits, customers, and ideal salesperson will come to nothing if you do not take the next most important step of the approach: Figure out how to match the type of product you have with the size of contracts you are hoping to sign. And, once you understand those deal parameters, you will need to hire for the type of salespeople best suited for delivering those results. The way to do that is to use what I call "the magic triangle."

This triangle represents the connection between your product, average sales price or contract, and your sales model. Your goal is to find the *right match between its three vertices* connecting those three elements and hence create your magic triangle, one that balances. This is easily drawn as follows:

Product
Definition

Average Deal Size   Sales Method

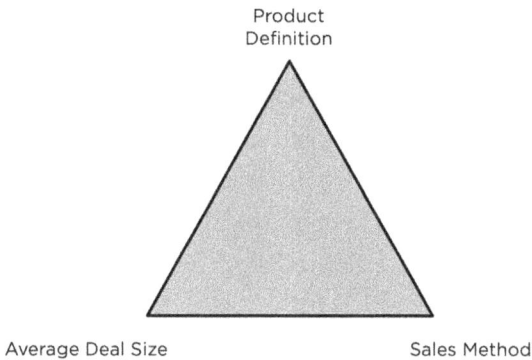

Here are the definitions of the vertices:

- **Product definition.** The product definition refers to the attributes in terms of complexity of your product and also specifically includes the friction involved in demonstrating your benefits and in integrating your product in the customer workflow. Some products require no permission or integration, some require a moderate amount, while others can involve a large set of connections, complex integrations, difficult-to-obtain permissions, and sometimes lengthy training on customer data or customization. The more complex your product is, the more time will be involved in your sales process, and sometimes time on-site will be required.

- **Average deal size.** The average deal size is simply the size of deal you'll aim to walk away with in the first engagement. Sometimes the initial deal is small but there is a high probability the customer will have wider use, and that will result in a much larger potential deal size which you need to account for. There may also be situations where customers want a multiyear deal, so you can think about the average deal size as the initial deal size plus a risk-adjusted evaluation of the total potential.

- **Sales method.** The sales method represents the type of sales approach required for your product and leads to a decision about the type of salesperson on the sales spectrum you deploy. The simplest approach starts with PLG or a simple one-touch phone call. The next levels would be multi-touch conducted by the transaction-focused salesperson, and the end of that spectrum is the long major-account sales cycle that involves working through the entire customer organization as conducted by senior sales executives.

To maximize the potential of your startup you will need to map your specific situation to your triangle, and there as many triangles as situations that fit every company. Every situation is different and the decision is based on all the work you've done up to this point on the Market Entry Strategy. How much friction is involved with selling your product? What size deal can you get? What sales methodology and types of salespeople will you employ? Once you have answers to those questions, place them into the triangle framework and see if it balances and makes sense. There is a continuous spectrum of possible triangles that balance based on different complexity of products, different deal sizes, and different sales methodologies. If you find that balance between those three elements, you have a magic triangle. But there are many cases where you'll discover that your triangle does not balance—a huge red flag for your ability to execute your Market Entry Strategy.

EXAMPLES OF MAGIC TRIANGLES

In this first triangle the product is straightforward—the prospective customer can install it without any assistance from your sales engineering team, they do not require any integration with any other product or database to obtain the benefits you are advertising, and the product is sufficiently self-explanatory that the customer can find the value in the POC by themselves. With those characteristics you can afford to build a large pipeline and a wide funnel because you have almost no cost of sales—the customer can understand your product, install it themselves, and derive the benefit. Given this market, in terms of price expectation and the ROI delivered by your product, you believe that these customers will agree to an initial purchase price between $500 and $5,000 and quite a few of them will simply charge it on their credit cards. If you encounter these characteristics for your product, you can expect to be able to sell in a single scripted phone call conducted by a junior salesperson. So, the sales model would either be PLG or a simple sales process.

A Simple Product/Sales Magic Triangle

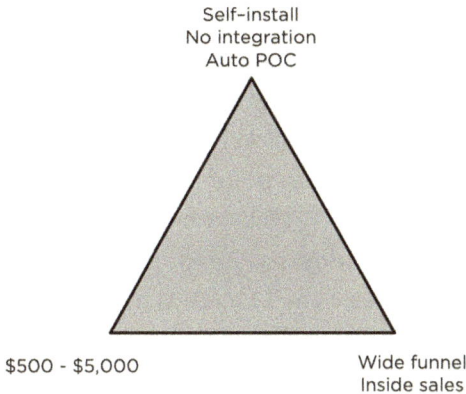

Self-install
No integration
Auto POC

$500 - $5,000

Wide funnel
Inside sales

This triangle balances. But recognize the requirements to make this work: the product must be extremely polished, and the value proposition must be obvious without anyone guiding the customer. There are lots of established products that fit this triangle, but very few new, innovative products are polished enough and have obvious value to the customer. Very few products genuinely have such a low amount of friction, and if you want to choose this magic triangle, you have to relentlessly strive for those characteristics. If your product does not clear that bar, customers will abandon your product or you will have to involve customer support in the sales process. If that happens, expect multiple phone calls from customers and your cost of sales will increase. If you only walk away with $5,000, you cannot afford any of these expenses to satisfy the customers and your market entry will imply a significant burn rate relative to the revenue level and other KPIs you'll be able to achieve.

I was on the board of a company that was adamant that the product was compatible with a PLG approach. This belief was informed in part by the fact that a previous product in an adjacent category had successfully grown to tens of millions in revenue using PLG—though it should be noted that they switched to a large-account sales motion thereafter. However, no matter how hard they tried, there was still some friction left. I tested that fact a few times by interviewing prospects I had direct access to and although they could download and install the product, they just couldn't insert it into their workflow and get to the value point on their own. The sales traction and conversion rates were frustratingly minimal. Eventually the company switched its sales process and started closing deals over $100,000 using a more classical transactional sales motion.

Let's look at an example of a product with medium complexity. In this second triangle we have a product that's more complex than the previous one. This product requires sales engineers to guide the customer through a POC, but it's a relatively short and straightforward process that can easily be executed over the phone. The customer has only simple integrations with another product or a database and can easily get to the point where they see the benefits demonstrated within their workflow. In this medium-complexity product, customers deal with only a small set of permissions required, which are relatively easy for them to obtain. Given this limited friction in the sales process, you expect to be using transaction-focused salespeople who can close the deal over the phone or after a single meeting with the customer. With these resources requirements—a transaction-focused salesperson who may or may not have to be on-site—every sales process now costs some real money. You'll want to be more targeted in your approach to customers to ensure a high conversion rate for the activity you will invest in. With this moderately complex product you hope to be able to demonstrate solid ROI, and the market expectation of price point for this type of product allows you to demand a first deal in the tens of thousands of dollars.

A Moderately Complex Product/Sales Magic Triangle

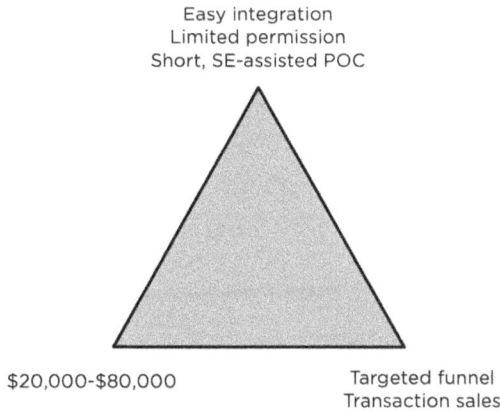

Easy integration
Limited permission
Short, SE-assisted POC

$20,000-$80,000   Targeted funnel
                                   Transaction sales

This magic triangle also balances: the customer needs some hands-on support but it can be managed by a transaction-focused salesperson. You'll target the customers who have the critical need for your product and can charge a reasonable price justified by the ROI. Where you may have trouble with this magic triangle is if you discover that there's considerably more friction to achieve the benefits than what you expected and your investment in time and resources escalates. You may also find that the customer qualification is more complex than what your salespeople can execute on or—as is often the case—your salespeople require more rigorous training and extreme discipline to achieve the conversion rate. You may also find that the ROI you anticipated may not be quite as compelling as you thought or that you are the trailblazer because this market has never valued this type of functionality at your price point and instead, the market has expectations of much lower price points. Any of these scenarios (and many others) will knock this triangle out of balance. Higher customer acquisition costs, lower conversion rates, or disappointing deal size leading to slow-growing top line will signal that your magic triangle is not balanced. The opposite also holds: if your product could command a higher price, you won't be able to obtain it because you have the wrong sales methodology. For a much higher-priced product, a transaction-focused salesperson will not have the gravitas (or skills) to be able to work with your customer's high-level executives. Your transaction-focused salespeople are not trained for that level of interaction and in all likelihood will be desperately afraid to ask for a larger deal. Your average deal size will be too low and, as a result, your triangle won't balance.

Let's now look at a product with high complexity. In this third triangle we have a seriously complicated product, one needing deep integration into a complex enterprise workflow. You should expect that the sales cycle will take months because you have to obtain all the permissions and connect your product to all the data sources and other products in the workflow. You might need to send a sales engineer on-site, sometimes over an extended period. However, given the significant business benefit you're delivering and the scale of the customer, those customers will be expected to sign a six-figure deal, perhaps even a seven-figure deal upfront or extended over some committed time frame. Executing on these types of engagements requires an understanding of the entire organization; it most likely requires C-level sponsorship and help to steer through all the barriers to complete a deal. Naturally, you recognize that selling a complex product to a large organization for six-to-seven figures will require a very experienced sales executive who can orchestrate the campaign and muster all the internal resources needed to delight this customer. But if this is your triangle, then the size of the deal is worth all that effort and a handful of deployments at customers will generate significant revenues and propel you to market leadership.

A Highly Complex Product/Sales Magic Triangle

Send sales engineer
Integrate with existing data
3-6 months POC

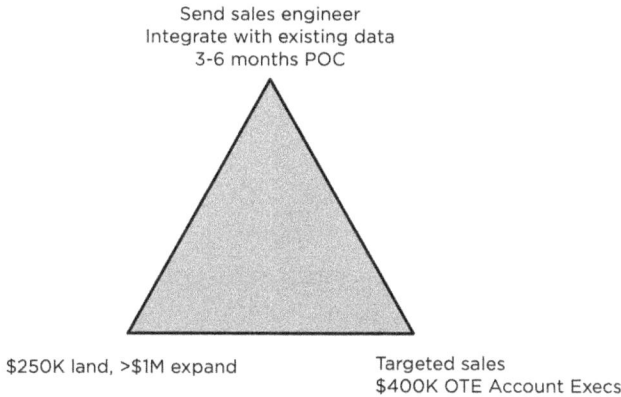

$250K land, >$1M expand   Targeted sales
                                           $400K OTE Account Execs

Again, this magic triangle balances, but it's critical to really understand the customer business and to have a solid case for the impact your
benefits deliver. If you fail to understand the customer deeply or fail to
show the business benefit, you may find out that you have a technical win
but not enough ROI for an executive to sponsor this contract. Also, if
you haven't aligned with the strategic direction of the customer through
a high-level vision match between your management team and theirs,
you'll get a much smaller deal or no deal at all, despite having spent frustrating months on one account. If your sales executive does not properly
map the account, that person may fail to recognize opposition or unseen
objections. If the sales executive overly relied on a champion, you may
think you have a deal but will realize too late that you do not.

YOUR FIRST SALES GOAL

The disciplined approach underlying the Market Entry Strategy applies to your sales model and requires that you find your magic triangle.
Finding the magic triangle that works for your startup and your product
is your first priority if you want to figure out the sales model for your
market entry. If you haven't figured out your magic triangle, you actually
do not know what kind of salespeople you need to hire and this is not

something you can afford to experiment with. Do not expect to hire a vice president of sales who's multitalented and can help you figure out what sales methodology will work for your product. You'll likely only be able to find a person who's aligned in character and experience with building a certain kind of organization that executes a specific type of sales process with a given kind of salesperson—and you can't expect them to discover the actual sales process you need and implement it. The sequence you should strive for is: first find your magic triangle, then hire the right salespeople for your product. Too often we see startups with a great product and a lot of upside that hire a top-notch salesperson only to flounder in the market because they skipped over the important step of finding their magic triangle and matching the right profile.

As we went through some examples of magic triangles and some of their failure modes, we've seen cases where the product does not align, others where the deal size does not meet your expectations, and yet others where your salespeople fail to match the sales process you require. It's critical to understand that you might need to *fix any of the vertices* of your triangle. Technical founding teams will often take the product as a fixed point in space—something that cannot be changed or modified—and they'll try to find a matching sales methodology. But if your market does not tolerate the price point that would balance the triangle with the complexity of your product, you need to change the product. For example, you might need to remove capabilities to make it simpler, so the customer journey is more straightforward. If you do that, you'll be able to balance a different magic triangle. Similarly, if your product is far more complex than what you expected but you find that larger enterprises are potential targets, then you can aim for a higher average deal size but will also need to hire much more senior executives than initially anticipated. The main lesson from the magic triangle is that you must balance your triangle. To do that, you need to be fully open-minded that any or several of your vertices might need to be reexamined. This isn't the time to say, "The product is the product," or, "The price is the price." If you do not have a magic triangle that's balanced, something has to change, or you will not succeed.

Getting to a balanced magic triangle requires a rigorous process of experimentation. In some markets, there are habits of customers that have been formed over decades and you won't be able to change them— you have to figure out how to work with those habits. If you have deep customer intimacy, discovering market habits shouldn't surprise you but you still need to do research to find out what those habits are. Regardless of what magic triangle you think is best, it's important to set up a clear plan to test it because you need to map your situation to your triangle. How complex is your product? What will be required to execute a POV and demonstrate your benefits in the customer's workflow? What is the sales process you will have to deploy to go from the first contact to the conclusion of a purchase? Will your product be a simple PLG, is it multi-touch, or does it require a complete account mapping? What is the business benefit ROI calculation you can put on the table to justify your price point? If you don't know the answers to these questions, you are flying blind and will need to create a clear set of actions that will deliver the data needed so you'll know where you stand.

When you run these experiments, and collect data, it's important to make sure you're not *selling* the customer. Your goal should be to openly communicate with your customer, to listen to them, not to convince them. For example, show the customer the business benefits and the impact you believe you have on their business—remember, you chose this customer and not some other customer because of your Market Entry Strategy. Provide the customer with estimates of what your product means for their ROI and ask them whether they agree with you. If not, how would they modify your assessments to conform to their reality? Similarly, explain how you expect your product to integrate in their workflow and ask them to verify your assumptions.

If the customer raises concerns about needing certain integrations and permissions, rather than convincing them those aren't needed, just listen. Do not be shy about asking about their organizational structure and what it would take to acquire your product. Who needs to approve, sponsor, or negotiate your product adoption? Who can block your product openly or behind the scenes? If you have customer intimacy you will get the answers to these questions, so make sure you listen carefully to each response and take in the data. If you have doubts about different triangles that are likely to balance, *experiment with each one deliberately and separately*. For example, try a certain number of mid-market accounts with a moderate complexity process AND try a set of enterprise accounts with a full sales approach but do so in a systematic and organized way. Or, if you really believe that your product doesn't need any sales support and that customers can figure out how to use it and derive the benefits on their own, watch your customers find your product, try it without any assistance, and then figure out the value that will make them produce their credit card. Often you'll see unexpected friction that will guide you to perhaps a different triangle.

YOUR TOP PRIORITY

Finding your magic triangle is priority #1 because without that, you cannot move forward in building a go-to-market organization. You won't be able to define the job descriptions of the people you need to hire without first understanding your magic triangle. And if you move forward anyway and find out down the road that you didn't balance your triangle, you most likely will have to terminate all the people you brought on board and invested in training and deploying because you hired the wrong people for a magic triangle that didn't work. Then, you'll have to hire a new set of people and hope you don't run out of money before demonstrating that this new triangle works. A typical startup raises money that will last eighteen months; if you fail in your first Market Entry Strategy, you will burn nine to twelve months of your runway. That is often fatal. If you're in the process of raising your early stage round, most investors will expect you to have a specific triangle in mind, to know how you will penetrate your market. Investors will want to know that you've figured out what you'll do with the money you will receive. Are you still experimenting or are you committed to a magic triangle? If you're still experimenting by the next round, it's unlikely you'll receive additional funding, so make finding your magic triangle your top priority.

EXERCISE

FIND YOUR MAGIC TRIANGLE

1. Rate your product complexity and its friction in integrating in the customer workflow and proving your business benefits.

2. Provide the range of price points for your product. Validate with the customer views of the ROI of your business benefits and the historical trends in your entry market.

3. Identify the type of salesperson you need for your magic triangle:

   - No salesperson / PLG

   - Remote scripted salesperson

   - Simple-process salesperson

   - Transaction-focused salesperson

   - Sales executive

4. Does this triangle balance? Would there be an alternative triangle that would also balance? If so, design a rigorous experiment to decide which one you will implement in your Market Entry Strategy.

KEY POINTS

- The magic triangle helps align the type of product you have with the expected contract size and the type of salesperson best suited for the associated sales process.

- Your magic triangle must balance to provide the types of deals you expect to close for your product in a cost-efficient way.

- Having an unbalanced triangle can lead to mispriced deals, unacceptable cost of sales, or inability to penetrate your entry market.

- Finding your magic triangle is your top priority and the foundation of building your sales team. Experiment rigorously and deliberately to find yours.

18

Scaling Sales

*"You can't run fast in water, and you
can't rapidly and efficiently scale sales."*

—JACQUES BENKOSKI, author

Have you ever tried running in water? It's easy to run fast in shallow water but if the water is knee-deep or deeper, it's exhausting, slow, and you
hardly make any progress compared to the energy expended. The reason?
Water has almost 1,000 times the drag of air and, everything else being
equal, the energy you need to move forward increases with the square
of your speed. So, as you try to accelerate in the water, you are facing an
impossible force of resistance and consequently, most people quickly get
exhausted trying. This analogy is useful because it translates to how a
startup should build a sales organization: if you do it at the right speed,
it is possible to do so efficiently. However, if you take a wrong approach
and attempt to do it faster, you'll waste a lot of energy, time, and money
for very little incremental progress.

WALKING IN WATER

If you have been applying the Market Entry Strategy advocated here, you
are now at the point where you have figured out your magic triangle.
With your magic triangle you know who you are selling to, how you are
selling to that customer, what average deal size you are targeting for your

product, and what type of salesperson you are looking to hire. So far you have either had founder-led sales, or sales involving the management team, or sometimes sales that required the entire company. But the first step of "walking in water" is to relinquish that sales role and have other people sell instead, as autonomously as possible. You now need to hire a small number of salespeople who fit your magic triangle and see if they can sell by themselves. Of course, just knowing your magic triangle isn't enough; you need to take the materials you created around company vision, mission, and positioning and develop those into sales-training materials. You then need to develop metrics to understand your effectiveness at ramping up people to be successful at sales in your entry market.

Depending on the level of complexity of your sales approach and the type of salespeople you decided to hire, you might need to have them understand the battle cards, the competitive analysis you conducted, and customers' expected objections. If you have a product that involves a POV, the salesperson will need to know when to bring that up with customers, how to set customer expectations, and how to manage those customers so you get a relevant POV. This is no small feat, but it's critical because that documentation, and the methods used to land customers, are essential to get to the point where you are confident that your sales methodology is working. Many teams skip entirely a rigorous sales-training approach and opt instead for a "training by osmosis" approach. This approach is based on the idea that people will learn by watching a startup founder or executive during the sales process and then imitate it. That might work but it does not scale; you won't be available for a larger number of people, and you are not setting the right foundations. In addition, it is an inappropriate experiment because your salespeople will not have your depth of knowledge and you won't develop the sales tools needed to train others in the company. It will appear as if the salespeople are making progress and learning your sales approach but as soon as you remove the safety net of your involvement, they are likely to fail. Instead, take the time to develop a training program that covers everything about your sales process and when your salespeople execute on that, and fully document it, you will then have a foundation from which to scale.

Your goal is to get to the point where this initial group of salespeople can sell on their own, in a repeatable sales process that aligns with the company's stated unique benefits. They need to be able to understand your use case, your differentiation, and all the background work you've done to successfully create your Market Entry Strategy, including your critical goal of getting to market leadership in your entry market. One problem I've seen a lot of startups make is focusing on the top-line revenue and tracking the sales team's bookings as the main metric. But top-line revenue is not by itself the key objective in your Market Entry Strategy. At the entry stage, consistency is your objective. If the sales team hits the bookings' goal, or even exceeds it, but the first customers are all different and do not match your entry market profile, you won't have the consistency to scale. Similarly, if the use cases are all over the place or if the average selling price is anywhere from $5,000 to $500,000, you will have failed at your market entry. This is not a scalable foundation, and each sale may be a one-off case for which you will struggle to find a common denominator.

A variation on this theme is companies that believe they can have more than one revenue stream. For example, a company might sell to one set of customers and monetize the data they have amassed with another type of player in their industry. Similarly in situations like two-sided marketplaces (where the startup bridges between two sets of customers wanting to connect), despite the irresistible urge to monetize both sides, this simply cannot be done at the market entry stage. You have to focus on *one* set of customers and *one* sales motion to have a chance at building a scalable sales process. The other revenue streams you might think about during the market entry stage should either be free (for example, in the two-sided marketplace) or simply not addressed at all. It's not even about whether it's strategically desirable, it's simply operationally impossible to be successful in your market entry if you have multiple revenue streams and hence multiple sets of customers. You need to focus on one single revenue stream until you have scaled your company, acquired an unassailable market leadership position, and developed, executed, and achieved stable, efficient processes. Then, maybe, very carefully you can experiment with a second revenue stream, applying a full Market Entry

Strategy approach to what is effectively a brand-new market entry stage for you. As you build out this new market entry, continue to monitor your position in your initial market and be prepared to refocus on that one at the slightest sign of weakening.

I've seen many examples of entrepreneurs pursuing more than one revenue stream; one example is startups selling into the US healthcare ecosystem. The healthcare market is complex and full of subtle motivations and interdependencies. Often, entrepreneurs will make a presentation that shows their startup attempting to monetize multiple avenues: self-insured enterprises, health-delivery organizations, health insurances, and sometimes even direct to the consumer. At the Series A, they will have obtained one to two customers in each revenue stream and show a minimal total revenue. In addition, the revenue model will be different for each revenue stream. The sales motion and sales materials will be matched to each case, and none of them will be sufficient proof of traction to justify an investment. Even if one of the streams is impressive, the other revenue streams will be seen as net negative to the investment thesis since they demonstrate a lack of focus and an organization that has already diverged.

A sophisticated early-stage investor will dig deeper into your revenue claims and analyze whether the organization you've put in place is executing on a cohesive plan. They will want to know, given additional capital, if your startup can scale in a clear way. Most importantly, if you have a scattered approach to customers in your entry market, you won't know how to build a scalable organization based on what you've started with because it isn't consistent. Even if you get additional funding, you will likely end up with a divergent organization or even a "spaghetti" where every salesperson ventures across the entire sales method spectrum. In this situation, while a salesperson may temporarily deliver the top-line expectations, the organization is doomed to failure or at the very least, to a deplorable capital efficiency.

Remember, your goal at the market entry stage is not only the revenue in the plan. Why? If you somehow achieve that but not the consistency of a sales methodology, sooner or later you'll hit a wall that's difficult to recover from. As you complete this stage, ask yourself: Do I have a foundation that will let me scale beyond this initial set of customers and with a larger sales team?

EXERCISE

WALKING IN WATER

After you find your magic triangle, answer each question.

1. Are your customer materials fully developed and ready to be handed to a sales team?

2. Are all your internal sales training materials fully developed? Start to train your first, small sales group (two to five people) according to a formal process (*not* by osmosis):

   - Prepare a formal training methodology and onboarding approach.

   - Validate your sales methodology with the newly hired sales team. Check if the observed magic triangle matches the expectation.

   - Validate again if your salespeople are on your positioning message, sell your unique business benefits, have coherent use cases and associated POV, and deliver expected deal sizes without your involvement.

KEY POINTS

- Scaling sales requires a balanced magic triangle where you know who you are selling to, how you are selling to that customer, the average deal size, and the type of salesperson needed.

- It is fundamental to create the underlying training materials to formally onboard new salespeople and move beyond the osmosis approach before you scale sales.

- Scaling sales can be done only if you have executed a Market Entry Strategy and can train salespeople on your entry market, your unique business benefits, and the critical customer need—as well as your differentiation, competitive landscape, and stated approach to market leadership.

- Top-line revenue is not by itself the key objective, consistency is your objective.

19

Bringing the Army to the Beach

"If you want to go fast, go alone.
If you want to go far, go together."

—AFRICAN PROVERB

If you've been implementing the Market Entry Strategy, you're now at the D-Day equivalent moment of "having landed on the beach." You've secured your initial leadership position in your entry market, identified the customers who have a problem that you're uniquely situated to solve, and created a small sales team trained to go out into the market and start selling. You may believe you're in a secure position, but nothing could be further from the truth. The next steps are just as critical as all the steps you've taken to reach this point. Unless you can bring in reinforcements, your beach conquest might be short-lived. If you start hiring a large number of people but don't have a clear plan of how to deploy them, you put both them and your startup in a perilous position. We see this situation in many startups; if they start to scale too early, they have to retreat and pivot—squandering resources, losing their market-leader position, and often having key contributors to the team quickly turn over.

"Landing on the beach" means you have successfully hired, trained, and proven that a small group of salespeople can sell on their own. Your next step is to bring your army to the beach, to hire more people and experiment to see if the foundation you laid out with training and deploying your initial sales force can scale. When you begin to hire the next group of salespeople, carefully observe and measure if they're ramping up as you were expecting. The larger the newly hired team is, and the further removed they are from the guidance and mentorship the executive group provided to your initial market entry sales team, the greater the chance that gaps or divergent approaches will appear. For example (and as previously mentioned), you may find that the positioning is drifting, or the use cases are widening for the sake of bringing in bookings—which salespeople are inherently wired to do. If either of these occur, you'll end up with a divergent organization. This is where you might find yourself listening to a recorded call with a prospect and not even recognize your own company in the way it's described and positioned by your own team. The worst thing you can do if this happens—if you start seeing some organizational drift—is to let it slide. Instead, take the time to retrench, retrain, and refocus on tighter sales management. Expect some friction and dissatisfaction within the sales team, but if you continue on the path of organizational drift, there is a steep price to be paid later.

The next question is, how fast can you really grow the organization? You have hired a second set of salespeople and feel you have put in place the materials and processes to bring them on board successfully, but how fast can you actually grow? The maximum sustainable growth rate will be directly related to the ramp-up time of salespeople and follows from simple math. For example, if you're attempting to grow your bookings by 100 percent year over year—a classical yardstick for startups—you will need to double the number of salespeople each year. More accurately, you'll need to double the number of "ramped" salespeople—people fully trained in all the nuances of your sales process, and this is where the ramp-up time comes into play.

Let's assume you're trying to double your sales team and look at the impact of taking three, six, or nine months to properly train and ramp a new hire in sales. If you were spending $100,000 on your sales team member as you started the year, you'll now spend $200,000. In the case of three months ramp-up time, $25,000 (.25x) of additional sales-related burn rate is spent without any corresponding revenue since your new hire salesperson is not actively selling at this point, but only training. If your ramp-up is six months, your burn will be $50,000 (.50x) and at nine months, it will be $75,000 (.75x). This is a conservative estimate of burn rate because for every salesperson you hire, there will be corresponding hires in sales engineering and lead generation. The amount of money you spend in ramp-up could easily be double or triple. You could save a lot of money if you had better training materials and onboarding processes and could reduce your ramp-up time. Keep in mind that not all sales-people will work out, regardless of how great you think your company is, and you should expect some churn, some turnover in the group. Each time you have churn and someone leaves, you are restarting the clock on wasted burn to train their replacement.

> **Unfortunately, I've been in situations where the ramp-up time majorly impacted the value of otherwise success-ful companies. In one case, the company had a complex product and never cracked the code for rapidly training salespeople on the positioning, unique business benefits, and the corresponding ideal use cases. Their training for a new salesperson was nine months and as salespeople struggled to reach their quotas, the company experi-enced an above-industry-average salesperson turnover. Even with high revenue levels, the startup simply could not afford to finance its growth. Why? The additional revenue generated didn't justify the investment in new sales hires. The implied increased burn rate didn't gener-ate an attractive revenue growth. We found ourselves in the unfortunate situation where raising more money to hire more salespeople was having a negative impact on the actual enterprise value.**

This is not a secondary issue if you want to grow your sales force. You need to think carefully about your ramp-up time and be honest with yourself about how long it actually takes to go from new hire to productive salesperson. Invest in training but also recognize the inherent friction that comes from your product complexity and your ability to execute on your positioning and also to teach market-leadership behavior. In some cases, you might even have to revisit your magic triangle because it doesn't actually balance at scale.

Another limiting factor that will impact the growth of the sales organization is related to human nature and the way we learn. There is a certain *reproduction speed*, which impacts every organization—the time it takes for someone to be hired, onboarded, become productive, and then become integrated enough in the organization that you can trust them to hire someone themselves. If you try to accelerate beyond your reproduction speed, you'll likely have unprepared staff hiring additional people. This is not a good situation and is almost guaranteed to lead to a tailspin in the quality of the organization, which translates into a degradation of all the metrics discussed so far. Breaking the maximum reproduction speed is akin to breaking through the sound barrier. Can you do it? Yes, but it requires a huge amount of energy, which creates a huge amount of vibration and which generates a big bang that has broken many airplanes.

Related to reproduction speed in hiring salespeople, there's an additional challenge from adding managerial layers in the sales organization. As your startup grows and you add more salespeople, you'll have to maintain proper sales processes, which requires a management layer. That managerial layer will provide regular oversight of individual salespeople. There's a limit to how many people one person can effectively manage, and most managers are unable to manage more than ten people (in the best case). You'll therefore need to add an intermediate layer of managers depending on your growth rate. This added layer of managers contributes to your expenses without bringing in a corresponding revenue line. Because of this added cost of an intermediate managerial layer, there's a natural tendency for startups to restrict growth in this layer, but the productivity of salespeople will degrade if they don't have the right structure. Weigh the pros and cons and the timing of adding sales management carefully and make sure to integrate that information in your plans.

One additional point about growth: keep in mind that when you're growing 2x year over year, 62 percent of your employees will have been with the company for less than 18 months. If you are growing 3x year over year, the number jumps to 78 percent. Yes, if you are growing that fast, nearly 80 percent of your workforce is relatively new. It is extraordinarily difficult to build a sustainable organization when the overwhelming majority of the team has just joined and you're tasking them to hire yet another set of employees. Last but not least, recognize that there's nothing more expensive than building an unsustainable organization. There is a frequently used mantra within venture capital of "triple-triple-double-double," referring to the goal of a startup to triple revenue the first two years and then double it the next two years. While that sounds really nice, it does not take into consideration the necessary questions of the sustainability and capital efficiency of a startup that's able to achieve that growth.

Let's say that you were able to grow your revenue fast and grow your sales force accordingly, regardless of the energy consumed. You would most likely eventually hit a wall and have to terminate a vast majority of the recent employees because you either mis-hired or mis-trained most of them, and then you'd have to hire a new set of employees. The cost to do this wholesale housecleaning and start over with a new slate of employees is far more than you'd expect—at least a significant portion, if perhaps not all, of your current funding round. And you would spend this money without having achieved the objectives that would unlock the next round because your revenue will not have grown in line with the amounts raised. So, be aware of the enticement of rapid growth—it might work out for you, but without paying attention to the pace suggested in the Market Entry Strategy, you're more likely to hire people who do not work out, you will have spiraling costs, and your valuation or even your survival will be affected.

EXERCISE

BRINGING YOUR ARMY TO THE BEACH

Once you are confident that your sales training and model are appropriate for your entry market, you'll need to figure out how fast you can scale.

1. How long does it take for a person to be hired by your organization, be trained, and become a fully functional salesperson?

2. Do you have all the metrics to measure this and the materials and processes to accelerate it?

3. What is the cost to your organization for each salesperson hired before they are productively selling? How much additional burn is spent in other departments?

4. What is the maximum "reproduction speed" in your organization?

5. At what point in your growth will it be necessary to add sales managers? Map that additional cost in your burn rate model.

KEY POINTS

- The basis of sales scaling is to create a small team of salespeople who can reliably sell in the market without management involvement.

- It is critical to onboard effectively and take into account the ramp-up time for new hires to become independent and productive. Your maximum sustainable growth rate is directly related to the ramp-up time.

- Scaling your startup as fast as possible is enticing but go too fast and the potential for organizational drift is high, undermining your Market Entry Strategy, creating an unsustainable burn rate, and negatively impacting your capital efficiency.

20

Strategic Sales Planning

"In preparing for battle, I have always found that plans are useless but planning is indispensable."

—GENERAL DWIGHT D. EISENHOWER,
Supreme Commander of the Allied Expeditionary Force

As you continue on your startup journey and begin to construct your sales force, using the Market Entry Strategy methodology implies that you develop a plan rather than trying a handful of random activities, and then trying to make sense of the feedback (based on the limited data you'll collect) in order to make far-ranging decisions. Instead by now, you'll have created a vision, a mission, and a company-level strategic plan that you translated into an operating plan. Fundamental to those four plans, you will have chosen an entry market and also have chosen to focus on one specific customer type, one geography, and selected other narrowing qualification criteria. By aligning all of these factors—vision, mission, strategic plan, operating plan, unique benefits, and entry market customers—you are setting up your startup to achieve early-market leadership. And as discussed earlier, being the market leader leads to a host of tangible benefits that will propel your startup and protect you from competitors.

The next step is to build a *strategic sales plan*, which addresses the problem of how you will contribute to the company's strategic and operating plans by executing a specific sales strategy. Different situations require different strategic sales plans, however. Just like there are many magic triangles, there are many strategic sales plans. Even for a given magic triangle, there's more than one strategy that could be successful. To build your plan, you are not starting from scratch, you can look to your completed magic triangle on one hand and to your strategic plan on the other hand. Logically, the strategic sales plan should bridge the two.

EXAMPLES OF STRATEGIC SALES PLANS

Let's examine a few examples that will introduce the concept and help you create your strategic sales plan. For example, if you have the magic triangle with the enterprise sales motion, you might decide to adopt a strategy that initially calls for pursuing "lighthouse customers" (i.e., customers commonly recognized as references for your market) for reputation purposes and to achieve market leadership by having public recognition rather than focusing on revenue.

Your strategic sales plan might look like this:

1. **Identify ten marquee targets in the US.** First, create a list of ten marquee (i.e., high profile) companies, those whose opinions have an influence in the industry you are focusing on for your entry market. Select them for that reputation over their revenue potential. But remember, you're also looking for the customers for whom your unique benefits will solve a critical need that's relevant to them and a use case you will be proud to showcase.

2. **Pursue these ten targets in a focused manner.** Next, pursue them at the exclusion of all others. You have selected them because of your Market Entry Strategy criteria, and they should be receptive or else they're not the right target—or you're not uniquely solving a critical need. For this targeting, perhaps you'll attend a trade show with the goal of meeting them specifically or use traditional

network connections to connect with their decision-maker. Regardless, the key objective is to convince maybe four out of the ten relevant marquee customers to move forward with proving your product superiority.

3. **Define the POV to match your spotlight differentiation.** Your third step is logically to prove that superiority as well as the differentiation of your product for your entry market. The way to achieve this goal is by following all the steps of the sale to demonstrate your unique business benefits to those customers. Those customers have the selected critical need and should want to engage to prove to themselves the value you claim to deliver. If that includes a POV, make sure the use case is indeed aligned with your Market Entry Strategy and that the customer will be comfortable sharing their results in a public success story or similar document.

4. **Drive usage first, revenue second.** The fourth step is to demonstrate your unique benefits in solving their critical need and use that to build credibility and cement your position as an early market leader. Remember for those initial marquee customers, your priority is to demonstrate your superiority and get to actual usage, even if the associated revenue is symbolic. Your win is the public white paper and the reference sale; the revenue growth will come from others.

5. **Enter the wider market with confirmed successes.** Finally, the fifth step in this sales strategy is to use the credibility you've acquired via the first steps to launch into the wider market. You are now in a position of having established that early leadership, which makes entry into the wider market easier because of the word-of-mouth resonance. In this last step, you can drive for efficient revenue growth, with the leverage of the proven benefits you've demonstrated.

This is a strategic sales plan that matches the magic triangle for this particular company, which is specific to its product, target customers, and average deal size during its market entry.

Let's look at another strategic sales plan for a company with a much simpler configuration. This magic triangle is based on a product that can be tested by the customer unassisted by your startup, you have both a free and a paid version of your product based on a certain usage model, and your sales team consists mostly of transaction-focused people. A company with these characteristics might have a strategic sales plan like this:

1. **Identify a vertical market segment.** Notice that in this case you are choosing your entry market, for example, perhaps an entire vertical segment, such as mid-market e-commerce companies.

2. **Systematically pursue one hundred names with account-based content marketing.** Given the narrowed vertical market segment, your goal is to find one hundred names in that entry market that care about your unique business benefits and for which you can easily demonstrate that you answer a critical need. These potential customers should be responsive to an account-based marketing approach using your clear positioning and messaging and recognize that you are solving something for them specifically. If they are not responsive, that should be a red flag about your positioning or messaging.

3. **Offer free version with auto-install.** Your goal is to get some initial adoption of your product by a small number of users within your entry market so you can see and validate that your product is solving a problem and getting usage. Monitor these early users to see if you are observing a natural expansion of usage as your benefits are demonstrated.

4. **Upsell 15 percent of top users to paid version using inside sales.** Given the simplicity of the install and expanding usage, you should be able to use inside sales to move the top users to a paid version based on your version of free-to-paid transition. Observe your metrics and demonstrate a scalable ratio of customer acquisition costs to lifetime value (CAC/LTV) based on that early land-and-expand motion.

5. **Expand to the rest of the mid-market e-commerce with confirmed successes.** Based on that initial success and proven sales model for your initial customers, you can scale the organization efficiently and address many more customers with identical use cases in the same industry. You should see signs of early market leadership with reference selling and adoption.

6. **At >$10 million in revenue, upscale to larger enterprises.** Having a solid leadership in your entry market and having improved the robustness and scalability of your product, experiment with larger enterprises with a higher initial deal size and more sophisticated salespeople.

This second strategic sales plan is very different from the first one because it has been adapted to this company's particular configuration of entry market, price point, and sales motion. Yet this second strategic sales plan is the one that fits this company's magic triangle; if you tried to switch the above plans and the companies, both would fail. And while there are many magic triangles that balance, there's an even larger number of strategic sales plans that can be created. The key takeaway? Don't leave your strategic sales plan to chance. You cannot go to your entry market and wing it; you must formulate a strategic sales plan that makes sense to your particular situation and is coherent with your magic triangle.

The advice for startups at this point in their journey is to take your time, get your team together, and think through all the elements of your magic triangle so that you create a strategic sales plan based on your configuration. Of course, this is not just an intellectual exercise but something that needs to be agreed to by all the stakeholders and aligned with the company's strategic plan and operation. The strategic sales plan will greatly impact your company at the market entry stage because it will align with and drive your marketing activity, impact your sales motion, determine your product direction, and especially dictate the type of salespeople you seek. It will also serve as a daily reminder of what your company is trying to achieve and will simplify your decision-making when distracting opportunities knock.

I was following a company that went to market without a strategic sales plan. They had figured out their magic triangle and understood very well what their unique business benefits were and how to demonstrate them. Yet they went haphazardly after any kind of customer, largely based on whom they could reach based on their personal and investors network. The first pass at going to market was a failure with low conversion rates, long sales cycles related to the average deal size, and mostly failing salespeople that would be let go after a few months. They reviewed the data, but couldn't pinpoint what was wrong. Was it the product, the sales execution, or the competitive position? The CEO went back to square one, letting go of most of the sales force and marketing staff, but designing a new and clear strategic sales plan. The target customers in this second plan were well defined and matched the startup's strategic sales goals. The CEO devised a dashboard that let them measure the marketing activity efficiency, and they honed in on the type of salespeople needed; their positioning then became clearer and the POVs more predictable. Did it all work magically? Not quite, but they could now at least reliably measure against their stated strategy to find where the gaps were and begin to incrementally improve. They eventually reached their expected KPIs, albeit after a costly round extension to make up for their early random experiment.

Is it possible (or even likely) that you get it wrong, that your strategic sales plan fails? Obviously, this does happen, and it actually happens more often than not. However, if you have a strategic sales plan, you can measure your progress (or lack of) against a baseline and recognize if you're advancing toward your objectives. You can measure the performance of your team and hold them accountable for their commitments. And if it doesn't work, you can think through which elements failed. Was it sales execution? Were there issues demonstrating the benefit value in your entry market? Did your product maturity cause sales friction? There are numerous valid hypotheses that you can test against your original assumptions, but as you have an actual strategic sales plan, you can do this testing in a much more transparent and clear way than using a scattered set of data points resulting from disjointed sales activities. Not only will you be able to generate hypotheses, but you'll also get to the answers much faster. With this new information it's much easier to replan because you can be assured that you have more reliable data. Then, when you design a new approach, you'll know how it differs from a previously well-understood alternative. Additionally, alignment won't be an issue because both the internal and external stakeholders will know where the company stands and will support the actions needed to increase your chances of having a second go before you run out of money.

The year-one goal for a startup is to find their magic triangle; this is the foundation for the first year of commercial activity. The year-two goal is to figure out your strategic sales plan because without it, you're likely going to struggle to align your organization and build that foundational layer for scaling. Although you can think of the strategic sales plan as a secondary part of your Market Entry Strategy, it is no less critical to your success. You need a plan, not just some ideas shot in all directions to see what works. We have seen countless examples of companies that have figured out their magic triangle and seemingly found product-market fit, yet failed in their next steps. Those companies reach $1 million in sales and hire a sales team but then fail to grow at a pace that matches their increased burn rate. What went wrong? They often didn't have a strategic sales plan, but instead had multiple customer profiles, divergent use cases, and no way to measure the deviations from their plan.

Last but not least, rest assured that most early-stage investors will want you to clearly understand your magic triangle, even though you may not have started to execute on it yet. However, they're more likely to back you if you have a compelling strategic sales plan and they can develop confidence that this is the plan you are expecting to execute with their money. At this early stage it's not expected that you have executed your strategic sales plan, or even proven that it works. However, showing the logic that led to the plan, the market you want to enter, the type of sales-people you want to hire, and the objectives you're setting for yourself—all of that can make a difference in whether you get funded. By the time you reach a growth stage round, though, you'll have to clear the bar with a proven working sales organization and initial sales efficiency metrics that can be measured against your strategic sales plan.

> One very telling sign that a company will not be able to scale is the spread in the quota achievement of their individual salespeople. You do not want a few top performers who are way over target and quite a few people who fail to make much progress. This is a telltale sign of a poorly crafted or executed strategic sales plan. Remember, you'll have to let these poor performers go, replace them, and then go through the expense of the ramp-up for the next set of replacements. You need to have a plan to measure performance against and to know why certain elements work while other elements don't work. If you haven't changed anything in your approach, the next set of new hires will likely have the same distribution. This will compound and lead to a much higher burn rate and a capital-inefficient or failed market entry.

EXERCISE

FIND YOUR STRATEGIC SALES PLAN

1. Identify your strategic sales objectives—what would success look like after two years?

2. Identify the approach you want to take to achieve your objectives: who would be your early customers?

3. What do you want to prove by winning them over (e.g., technical proof, sales motion efficiency, usage, revenue model, etc.)?

4. What type of customers would match those goals and what approach will you take to get to them?

5. List the initial customers you'll systematically pursue as part of your strategic sales plan.

6. Develop metrics for success, including intermediate indications before you actually see revenue growth.

KEY POINTS

- A strategic sales plan needs to be designed and aligned with the company's strategic and operating plans and work with their magic triangle.

- There are many possible strategic sales plans. You might not get it completely right but it's critical to have one in order to align the organization behind clear goals.

- Metrics for strategic sales plan success will give you early indications if you are ready to scale your sales organization.

PART V

WINNING YOUR MARKET ENTRY

21

Strategy
to Execution

*"Strategy equals execution. All the great
ideas and visions in the world are worthless if they
can't be implemented rapidly and efficiently.
Good leaders delegate and empower others liberally,
but they pay attention to details, every day.*

—COLIN POWELL, Former Secretary of State

This quote by Colin Powell comes closest to expressing the nuances of strategy, execution, and leadership because it embodies the tension between delegating to others yet paying attention to details and framing that within a context we all understand: the daily grind of getting things done. In the specific context of the Market Entry Strategy, we need to understand how execution and delegation are going through an important evolution while the startup is going through its market entry stage.

Since the earliest days of your startup, you invested most of your time and resources into defining and developing your product. Ideally, you also started thinking about your market entry, including your spotlight, your differentiation, your competitive position, and your sales strategy. But your efforts so far have been internally focused; now that you're graduating through the market entry stage, you need to focus externally. This new mindset from internal to external initiates a transformation that is critical but often overlooked: the CEO and the management team need to change how they operate *from doing everything to doing nothing.*

This may seem shocking at first, and it does not mean literally "doing nothing." It doesn't mean that the CEO and management team can collect a paycheck while spending time on the golf course waiting for a unicorn exit. It means that, in contrast to the early days of a startup when the founding team was busy with many menial tasks that required a lot of their time, things that were not yet possible to delegate to others, this new phase is different. The early-stage tasks typically include registering the company, crafting an initial fundraising presentation, and many more mundane tasks such as finding an office and signing a lease, registering the company domain, buying computers, desks, or a coffee machine. In the initial days of a new company, these activities are exciting because they are the visual marks of the existence of the new entity.

Before your market entry stage, it is also completely appropriate and expected for the CEO to personally review marketing documents and success stories or provide opinions on user interface mock-ups. It's common for CEOs to use the product themselves to get a clear sense of the workflow and a better understanding of its customer readiness. Similarly, the CEO and executive team are directly involved in initial customer contacts, and with relationship building with analysts and other influential people in the industry. Everyone on the startup's initial team is doing a little bit of everything, and these initial high-intensity interactions are critical foundations of the company.

However, once you get into the market entry stage, you need to expand your management team, hire salespeople and sales engineers. You need to build a customer-success team that can deliver a proper support plan that separates engineering and development from the continuous interruptions from early customers. In this second phase, the CEO needs an entirely new set of priorities so they can build the foundation of company scaling. These new priorities run parallel to the market entry stage—they're not something you wait to consider after you complete your market entry. This second stage is characterized by a change from everyone rolling up their sleeves and putting out fires as they appear, to a management approach where everyone owns their domain and is held responsible for deliverables. The job description of a CEO in this second phase is no longer to play whack-a-mole, to be highly immersed in every detail, or to be the arbitrator of every decision, as they have been up to this point. If you want to scale and succeed after your market entry stage, the role of the CEO needs to change from doing everything to doing nothing and trusting the people you've put in place to do everything.

THE REAL CEO JOB

So, the job the CEO did until the market entry stage was not really the CEO job. Yes, you had the title "CEO" but were in a temporary stage where the urgency of the day dictated what you were doing. *That is not the CEO job.* The actual CEO job description is contained in the early chapters of this book: Your job as the CEO is first and foremost to articulate a vision—once you have that, your job is to describe the company's mission, to create a strategic plan, and to execute on the operating plan. That is the actual number one responsibility of the CEO: *to build a plan for the company to execute.* It is difficult to first come up with these interdependent plans and doubly difficult to then obtain the buy-in of the executive team and constantly maintain the alignment internally and externally. This is your full-time job.

A key element of this alignment is maintaining discipline about your messaging and positioning. Earlier we stated that this is the foundation of everything. It may sound simple and trivial, but it is a daily responsibility that rests on your shoulders. When you meet with investors and analysts, when you are being quoted in press releases and funding announcements, you need to be vigilant and remember that to be successful, you need one, and only one, spotlight. That discipline applies equally to your interactions with your internal stakeholders. Your challenge is to have every employee understanding the vision-mission-strategic plan-operating plan so well that they can recite them verbatim. Each employee needs to know where the company is headed at a high level, but also where their work and efforts fit into that bigger picture so that they can be empowered to make daily decisions with increasing autonomy and enable the delegation that you need to achieve in order to scale.

Early in my career, I had the opportunity to travel with the CEO of a 17,000-employee company and at every company site we visited—and to every employee we met—the CEO repeated the same vision and mission and plans dozens of times each day—to the word—as if they were saying it the first time that day. This is the grind of leadership, but great leaders master the art of doing that because they know that the never-ending communication of the company strategy is the cornerstone to keeping every employee aligned and motivated.

Even if you achieve broad internal support for the vision-mission-strategic-operating plans, it is still a continuous effort by the CEO to reinforce those plans, which will occasionally become frustratingly repetitive. I guarantee that nearly every time you have an all-hands meeting and explain the tenets of your Market Entry Strategy (as you always should), at least one employee at the end of the session will raise their hand and ask a question as if they hadn't just heard you explain the company's plans. *Every time.* Don't be discouraged, just repeat them.

EXERCISE

THE REAL CEO JOB

1. What portion of your time do you spend on the day-to-day operations?

2. Did you develop vision, mission, strategic plan, and operating plans and do you feel ownership of them?

3. Did you get full ownership from your executives and feel they can repeat the vision, mission, strategic, and operating plans?

4. Are you repeating and evangelizing the mission, vision, and the strategic and operating objectives in every all-hands meeting?

5. Are you repeating the vision and mission in every external meeting?

6. During the past week, how many times have you done the above? (*Hint:* Five is a minimum, ten is okay, above ten is better.)

7. Look at your calendar and make a pie chart of where you spent your time last week to check yourself on whether you're doing the real CEO job.

BUILDING THE PLAN

If the key role of the CEO is to build a plan, how do you do that? One of the best ways I have experienced to build a plan is based on the principle of *backward planning*. The mainstay of the backward-planning method is to first imagine your company sometime in the future—if you're at market entry, a good time frame is eighteen to twenty-four months (six to eight quarters). For startups, it often coincides with what you'll need to achieve before you start your next round of financing, and eighteen months is a typical time frame for that.

Next, assemble your executive team in a room or at an off-site location because the "building a plan" exercise requires their undivided attention. There, each person contributes to imagining that future company, and they should work collectively to be as detailed and exhaustive in describing it as possible. Some examples of the key indicators you will try to imagine for your ideal future company are:

- What is its revenue run rate?

- How many customers does the company have?

- How many sales engagements?

- What is the total value of the qualified pipeline?

- How many employees?

- How many people are in each function—e.g. sales/support/marketing/engineering/general and administrative (G&A)?

- What is the burn rate?

- What are the quarterly collections?

- What is the churn rate?

- How many offices?

This is a very superficial list to give you a sense of the granularity. It's critical to dig as deep as possible and describe that ideal company in all its components.

> If you're going to run out of money earlier unless you get more financing, then any milestone you put into your plan that would have an impact after you run out of money needs to be set aside. Any milestone after the running-out-of-money point is irrelevant because all of the resources at your disposal need to be focused on what will let you stay alive! That doesn't mean that you do not embark on long-term projects, but it does mean that any long-term project impact must be measurable, and therefore "diligenced" within the required time frame. For example, you might not have completed the migration of all your customers to a new and much faster platform, but some have been moved and demonstrated the benefits, so such a migration can still be in your plan.

Once you have this broad, exhaustive, detailed list of what your future company will look like in eighteen months, iterate with your team on this until you have something solid and that you all agree on. Test whether this future company will meet the requirements for your next funding round by talking to current and friendly future investors. Compare your plan to available industry data for productivity, conversion rates, salaries, and other costs to make sure your assumptions are realistic. Verify again and again that all your executives agree with your future company and if you sense any discomfort in the room, pull on that string until you've discovered and addressed any element that someone might think is not consistent or possible to achieve.

Here is where the backward planning comes into play because you now have an ideal future company you have imagined with a list of characteristics and future KPIs that you and your executive team agree on. Your next step is to look at what needs to happen a quarter *before* this ideal future. For example, if you imagined that your company would have 50 customers in 6 quarters, how many do you need to have before the beginning of that quarter? Maybe you expect to have 40 customers, so adding 10 customers will be your goal for that 6th quarter. You might next ask: For those additional 10 customers to be closed in that quarter:

- How many active sales engagements do we need?

- How many simultaneous POVs do we have to run?

- How many sales engineers do those require?

- How many support people do we need?

- What lead generation activity do we need to complete the quarter before?

- What marketing staff and campaign expenses do we need to budget for?

- To get the staff we need to hire, how many new-candidate leads do we need to have, and how much time will be spent interviewing them?

These are just a sample of the questions you will have to ask and answer; you can imagine there are many more that will be required to look at every element of the company that quarter.

So, you will need to analyze all that data and assess what has to happen across all the same parameters before the end quarter you're looking at. If you repeat this exercise going backward through each quarter all the way to today, you will find all the goals that you need to achieve to match the assumptions you made for the ideal company you envisioned 18 months from now. By building the plan this way, every set of quarterly goals will determine the previous quarter goals and reveal the required associated activity in every quarter.

Backward-Planning Method

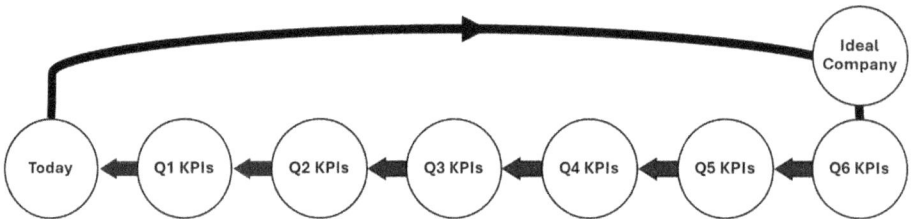

You might already have some misgivings about your plans. For example, in thinking about the 50 customers needed by quarter 6, you might have trepidations about how to generate the leads needed that will convert to deals and yield the 10 new customers you expect. To complete the picture, imagine that to generate the leads needed to get to 50 customers, you might need an increased marketing budget, with staff and program expenses ballooning. If that happens, your burn goes up and your runway shrinks. And as you keep going back, you might discover that to achieve your plan, what you need to accomplish already in this current quarter and the next one is just beyond the scaling capacity you thought you could execute on. Or, as mentioned in the sales-scaling discussion, as you hire new salespeople and train them, you are increasing your expenses without generating revenue. Because for a given burn rate, you can hire only so many new salespeople, this in turn means you can only grow your revenue at a certain pace. By using a backward-planning method, all these issues which arise in the course of scaling can be analyzed and understood *before* you reach those decisive moments.

What do you do if the future company in your thought experiment simply seems impossible? You iterate and look to see where things need to change. You might need more money than you thought, or maybe that eighteen-month imaginary company needs to be revised so that you grow more slowly than you originally imagined. Maybe you'll need to change priorities because you recognize where your most critical hires and budgets must be allocated to increase the chances of achieving a given milestone.

This is a point where, as CEO, you need to engage your team fully and make sure all their fears, all their misgivings, are brought to the surface so that any disconnection in the KPIs can be uncovered. Keep iterating until you come up with a set of quarterly objectives for each department that the executive in charge has participated in elaborating. A classic example is that the head of sales won't be able to fully agree to a revenue target unless the head of marketing can deliver the leads and the head of product can deliver a product that matches the type of customers you expect to close at that time. While the KPIs you come up with are important, the collaboration process you undertake to reach that alignment is critical.

Although this method seems abstract and maybe even difficult to comprehend until you try it, it is a proven way to create an operational plan that aligns everyone in the company to the same metrics and milestones. My experience has taught me that committing to the process, from spending time with your executive team doing a future-looking thought experiment, to iterating on that and finally creating a plan that everybody believes in, will produce spectacular results. The hours you spend working on your plan will be paid back hundreds of times over.

One last positive by-product is that your plan will also serve as a foundation for your communication with the board. Start every board meeting by reminding everyone of the strategic and operating plans you are all aligned with, then provide a simple scorecard showing your quarterly progress against the agreed KPIs since the last board meeting. This way, the board meetings will serve as a metronome that will rhythm the progress of the company toward its objectives and a time and place where every executive is publicly accountable for what they have committed to in the plan.

EXERCISE

YOUR EIGHTEEN-TO-TWENTY-FOUR-MONTH PLAN

1. Build a written description of your company in eighteen to twenty-four months with all the associated KPIs.

2. Consider the previous quarter and confirm that this quarter's KPIs could be achieved and iron out the differences.

3. Walk back recursively all the way to this current quarter using the methodology.

4. What mismatches did you discover? How did you adjust the KPIs to make the plan coherent?

5. Were all your executives involved, have they expressed their opinions, and have they willingly bought into the plan?

6. Have you presented your plan to your board?

7. Have you checked your plan with possible next-round investors?

8. Are you committed to showing the plan at all-hands meetings?

9. Are you ready to hold your teams accountable for the plan all of you have built together?

THE OPERATING PLAN IS THE CORNERSTONE OF EXECUTION

The key to executing on the strategic and operating plans is to get the executives to buy into them, own them, and agree to be held accountable for the goals they have willingly signed up for. And if you have executed on the previous steps of the methodology, it ought to be easy to get that agreement since they've been involved in creating those plans since their inception. If you didn't do those exercises, or if you did them yourself and didn't include your team, you have a steeper hill to climb and everything will be more frustrating: getting buy-in, aligning and motivating people, and delivering results.

As CEO your highest priority is executing the plan, which means hiring, holding accountable—and if needed, firing—people to ensure that your plan comes to fruition. If you see mishaps or defective quality in any deliverable, it is essential that you resist the urge to return to your early startup CEO behavior, roll up your sleeves, and get involved in fixing the concern. The minute you dive in and participate in executive team members' issues, you lose your ability to hold those people accountable. You may think you're helping, but you are actually making things worse because once you make yourself part of the solution, you now also share in the ownership of the outcome. By stepping in, you are enabling that person to believe you'll be there to help with whatever milestone or task they're working on, and they are released from the responsibility of achieving their goals. As easy as it is for you to solve their problem, as CEO you need to remember that doing nothing is your job.

As an added incentive to focus on executing the plan, keep in mind that all the other executives, and often the entire company, will be watching the burgeoning culture and will see you tolerating the lack of performance. The message you're sending—every time you help or manage a situation—is that if a person does not deliver, they can always draft you into helping solve the problem and absolve themselves from responsibility. In the worst case (which happens surprisingly often), the habit that develops is that there's no point in trying to fully complete the tasks since the CEO will dive in and take charge anyway. Companies with this culture revert to the early startup behavior and never scale. They're dysfunctional, and the CEO, dealing with all the minutiae 24x7, becomes deeply frustrated while also being perceived as a micromanager.

It's important to put the changed role of CEO into perspective; I'm not advocating that you never help others on the team. Good business practice suggests that the CEO should work with others in the company by asking, "How can I help you?" I agree with that, but what a CEO needs to do is make available the necessary resources, provide an environment that is supportive of the individual, and build a framework for making autonomous decisions. That framework is grounded in the Market Entry Strategy and all that's been discussed so far. The operating plan you have developed with your team becomes a sacred pact that underpins the company's structure, supports its performance, and allows the CEO to focus on their real job.

NO EXECUTION WITHOUT ALIGNMENT

As the company grows, the executive team will naturally get larger with the hiring of new vice presidents who own their specific domain. Recognize that the cohesion of the executive team—new and old—and their buy-in to the operational plan you collectively created is key. When you interview candidates for these senior-level positions, it is vital that, in addition to their knowledge of the domain they'll be responsible for, you also understand their ability to contribute to the executive team. It's just as important to build a cohesive executive team as it is to hire champions in their field. That does not mean that every new hire will abdicate their opinions and abide by plans that may have been set before they joined, but to the contrary, they need to be tested for their ability and willingness to participate in your planning process. The operational plan will need to be modified, and so time and again you will assemble the executive team, look ahead several quarters, and take corrective action; you want to imagine how the group will work cohesively in that future setting. Every executive needs to be able to speak their mind and contribute to the discussions. It would be a shame if you could not fully benefit from the input of the more recently hired—especially as your new team members will typically bring a wealth of experience in the field and knowledge of how the company will need to operate at the next stage.

Sometimes it's possible for a CEO to accept that people disagree with a decision, but still commit to execute. What you cannot tolerate is an executive who comes out of the room seemingly in agreement but then decides to deride the decision in private, or even execute within their department to a different set of objectives. You will have a dysfunctional organization if the executive team is not aligned, if they are not in agreement on the plan moving forward.

> **There is often a debate about whether companies with founder-led teams have superiority over those with hired management replacing them. I'm in the camp that believes that if the founders can scale, the company is stronger for keeping them in their position because of their wealth of historical knowledge and singular ability to motivate the team. The caveat is that they have to scale. It is critical that people be judged on their ability to do their job at the time, regardless of their participation in the early days of the company. Executives hired later in the life of a company need to feel that they are equal members of the management team, and not second-class citizens to the founders who somehow enjoy membership in an inner circle or immunity from performance.**

Because executive alignment is so critical, a good practice is to review all the members of your executive team periodically for alignment and to measure their performance in that respect, just as much as you would review their individual operational results. Review also your hires a few months after they join and ask yourself: "Would I still hire them, knowing what I now know about them?" An important part of the real CEO job is to be continuously working to integrate your executives into a cohesive team, be ready to assess their input about a possible change of plan, but in the meantime continue enforcing execution on the current plan. As the owner of the plans, the CEO must constantly ensure that alignment and the execution of the plan.

EXERCISE

IS YOUR TEAM REALLY ALIGNED?

1. Do all your executives accept the commitments they are making as part of the plan? Are they willing to be held publicly accountable in staff meetings, board meetings, and at all-hands meetings?

2. Grade each of your executives on:
   - Commitment to the plan
   - Owning their objectives and willingness to be publicly accountable for them
   - Ability to communicate the plan to their teams
   - Collaboratively working to remedy issues or openly replan

REPLANNING IS NOT FAILURE

This planning methodology is powerful in itself, but in conjunction with the ideas of *the other side of focus*—the "no" list you created detailing opportunities you will not pursue and the divergent organization that you will avoid—your operational plan allows you to keep the company steered in the right direction. Executing your Market Entry Strategy means recognizing that as a startup, the environment you face is dynamic, and it can seem that six quarters into the future is an eternity. How do you react when new information surfaces? What happens if a change in the competitive landscape reduces your uniqueness, or if your magic triangle didn't quite fit as beautifully as you imagined? What do you do if there's more friction or lower average selling price than you anticipated? I highly recommend that you recognize those situations and address them immediately.

The common mistake is to take in the new data but then rather than replan, use that information instead to justify letting the operating plan drift. In accepting your operating plan drift, you progressively (and often much faster than you imagined) end up with a plan that's no longer relevant. You have KPIs that do not measure anything valuable, and executives have plenty of excuses to avoid being held accountable for results. Instead, take the time to repeat the process of creating your operational plan, again setting objectives and walking back, quarter by quarter. Take a fresh look at what requirements need to be achieved to make this new operational plan a reality. Just like the first time you created the plan, a key goal is to make sure that all executives are aligned and committed to this new future. The replanning process will go faster since it is (hopefully) an incremental change. Regardless, it's invaluable to equip the company with a new set of realistic objectives to strive for within the time frame left in your eighteen-month plan. A best practice is to keep the operating plan updated on a rolling six-quarter basis so that the company always has clear directions it knows to follow.

There is a tendency for CEOs to try to maintain the illusion that they can still achieve their original plan and to state this in front of key stakeholders such as the board and the employees. Often, the CEO's fear is that when either constituency recognizes the gap, it will create a backlash. And so this illusion is perpetuated. What if, for example, the board thinks the CEO and the management team can't execute? What will happen if the internal teams learn the truth? Will they become disappointed and demotivated by the slippage? Will you see an increase in the departure of key contributors? In my experience, those fears are unfounded and the reality is often the complete opposite. Both the board and the internal teams will soon discover what's going on anyway. They will figure out that there's been some drift in the operational plan and they might even identify the drift before you do. The board will realize it because they know how to pattern-match to certain missed KPIs, while the internal teams will know it because they're facing the frontline reality every day—and the rumor mill will take care of connecting the dots.

Instead, address your issues head-on and replan. The board will trust you because they see that you're realistic and have a new plan that's probably better than any alternative they can think of. The employees will trust you because they'll see a leader who is honest. Most importantly, employees are more likely to stick around if there is a reset plan that they believe in over a plan they know is unrealistic and that perpetuates a confidence gap with the management team. Have a single honest and achievable plan, a *single story* that you tell all the internal and external stakeholders. There is only one narrative: the true one.

One of the surprising facts I have observed time and again with all the companies I've ever been involved with is the significant impact of an on-plan culture versus an optimistic but unrealistic plan. Some companies have a culture where they beat or exceed the plan and deliver the metrics quarter after quarter. Other companies consistently create overly optimistic plans and come up short every time. Typically, behind that culture is a CEO or an executive team behavior that subscribes to a management perspective that you can better motivate people by giving them stretch goals (which they might achieve if all the moons align). "Stretch goals" sound great, but are useless if nobody seriously thinks they are possible. Replanning when required creates a winning attitude that is more potent in engaging people to push their boundaries than unrealistic stretch goals. This also gives you tailwinds with your board and your current investors, and you can show up to your next meeting proving that you delivered on the objectives you committed to.

As we conclude this last chapter, I recognize that this evolution of the CEO and the management team through their market entry phase is both critical to the startup's success and incredibly difficult to execute. It's easier to change systems and processes than to change people, their characters, and their habits. Yet, as an advisor to startups, I find it fascinating to watch and help CEOs and management teams grow as individuals, both professionally and personally, through this stage. It has been the source of my greatest satisfaction as an entrepreneur, investor, and board member. The elements in this chapter are not easy to execute but they represent a system that's been proven effective, and even when it fails, it gives a framework to analyze the situation and bounce back.

I was on the board of a company that was on a rapid growth curve and was quickly becoming a market leader in its entry market with the associated signs of lowered sales costs and capital efficiency. As is often the case, the company got an early acquisition offer. It was below the potential of the company but it was a meaningful life event for the founders. Through many discussions, the board, with the full support of the founders, rejected the offer and went back to work on growing the company. Two years later, the company was acquired for four times the original amount. When I met the CEO after the transaction closed, that person said: "Even if the final price would have been the same as the first offer, I would not have regretted those last two years because of what I learned and how much I grew as an individual."

KEY POINTS

- As you grow during the market entry stage, the role of the CEO shifts from day-to-day minutiae to building a foundation for scaling: the real job of the CEO is to build a plan for the company to execute.

- Build your plan using backward planning, imagining your company eighteen to twenty-four months out and recursively asking what needs to be accomplished in each prior quarter to reach the associated KPIs.

- Executing the company plan is enabled by gaining alignment from all stakeholders, hiring the right executive team, and holding them accountable for their commitments.

- Start every board meeting, staff meeting, and all-hands meeting by reminding everyone of the strategic and operating plans and the current agreed KPIs the company is pursuing.

- As new information becomes available, do not ignore or attempt to fit new data to your plan because this will lead to organizational drift. Instead, repeat the planning process and create a new plan.

- Strive to build a company where realistic objectives are achieved and transparency is celebrated. Have only one single narrative for all stakeholders.

- Recognize the challenges and the personal growth required of the CEO and management team to succeed during the market entry phase.

22

Conclusion

*"The most difficult thing is the decision to act,
the rest is merely tenacity."*

—AMELIA EARHART, aviation pioneer

Over the last two decades, we have seen an explosion in the number of startups and the associated capital available to them, as well as increased interest by the general public. Television shows like *Shark Tank*, for instance, have popularized some of the startup concepts and brought a layer of glamor to entrepreneurship, but at the cost of simplification for the sake of vulgarization. One consequence of this widespread interest in entrepreneurship is that people have applied the model of venture-backed startup to just about any domain—from mattress manufacturing to temporary office rentals—regardless of whether or not those domains fit the narrow venture model. We are in an age where the underlying principles of venture-model logic have been ignored: finance an initial effort, usually for a differentiated breakthrough, that then yields a sustainably superior product that enables building a leading company through focused execution which then allows it to pay back the original investment many times over. Associated with that is a Darwinian principle, where each round of financing leads to the next if the company executes or the financing dries up if it doesn't. Only a small number of startups will survive and take the leading position that provides significant returns to investors.

As the startup ecosystem has evolved and dramatically widened, it is now clear that the old market entry models need rethinking or at least updating. The emphasis in startup ecosystems needs to shift from market penetration through iterations on a wide market to focus instead on *differentiation* as the foundation to a narrower entry market to ensure a leadership position. Based on my experience, I realized that to close the gap, it was necessary to provide a rigorous framework to guide startups through their initial stages as a departure from both the random experimenting as well as an alternative to the unrealistic glorification of the invulnerable entrepreneur who succeeds through sheer perseverance. The growing acknowledgment of the mental cost of the current models used widely by the startup ecosystem has made it all the more motivating to propose an alternative.

Entrepreneurship is messy, ambiguous, and rife with unknowns about the future and what those might mean for a startup's success. The Market Entry Strategy emphasizes taking back control of all of the decisions a startup faces and is a reference framework that, perhaps imperfectly, provides a formal methodology to the critical market entry stage. It allows entrepreneurs and executive teams to know where they stand in their journey and it provides a method to develop a clear plan they can build their company around. Recognizing that the path they are on is failing and enabling them to take corrective actions is critical, even if that means going back to the drawing board because that can help them avoid spending years of their lives wastefully. Needless to say, providing the CEO and executive teams with a way to build strategic and operating plans to develop a scalable model allows them to move away from the mentally costly directionless daily grind and contributes to their personal growth.

The Market Entry Strategy helps in providing startups and executives with a clear series of steps to take in creating a business from its first day until you achieve leadership in your entry market. The power of the approach derives from actually thinking carefully and completing the associated exercises. In doing each exercise you will not only set up your startup to attack your market entry but will also gain one of the most important benefits: deep and broad alignment on what you are doing, why you are doing what you're doing, and what your short- and long-term plans are. I can't stress enough how important alignment is. At the very least, being in complete alignment saves you time and for a startup, time is the critical constraint. By being in alignment you increase your capital efficiency and your runway since you will not diverge from your critical focus; you won't have to struggle with opportunities that emerge because you have already completed and agreed to your "no" list. You will know which customers to approach, why you are approaching those customers, what benefits your startup provides, what sales model to use, and what salespeople to hire. Many of the critical decisions that most startups iterate and experiment with will be brought back into your hands and will be your decision to make because the Market Entry Strategy insists that you think about, discuss, and agree on all of these elements before you launch into action.

I have found, in working with multiple startups over the past several decades, that the Market Entry Strategy can have a tangible impact on a startup's outcome. A few years ago I had an unusual coincidence of events and ended up with seven M&A exits within a single year. As you would expect, some companies were sold below the invested value, some were solid returners that were impacted by the capital intensity of their imperfect market entry, and some companies achieved exceptional outcomes. As the year ended, I looked back on the seven board meetings in which we considered acquisition offers and reviewed the decisions we made. Though the situations were all over the spectrum, the common thread was logically whether the board believed that continuing to build would provide a better risk-adjusted outcome. And each time, the deciding factor in those discussions was the existence of a solid plan that was bought into by the entire management team and the board. In addition, having a solid plan allowed the M&A negotiators to have a legitimate claim that, unless the offer was attractive, the company had a plan-of-record that it would execute on and build more value. In itself, these two elements strongly justify having a strong execution methodology such as advocated in the Market Entry Strategy.

One final thought is this personal note: I have been an amateur competitive sailor for many years, and I started this book with what I thought was a judicious quote about sailing from Diane Green. I too have come to believe that there are strong parallels and lessons between sailing and startups, and I want to take a moment to share them with you. If you've never witnessed the start of a sailing race, it's important to understand that the beginning is critical. The starting line consists of an imaginary, invisible line that goes from the committee boat to a buoy. The boats line up near the line, circling each other to try to get the best position. Then the gun goes off and all boats try to cross the line in the front. This is not just a temporary advantage but one that will have a huge impact on the outcome of the regatta. The lead boat gets advantages the other boats don't get—clear air, leaving the competitors to deal with the disturbed flow created by its sails. The lead boat can also control the maneuvers of the others, forcing them to turn when they don't want to, or blocking them from turning when they want. The lead boat can cover any move others make, staying squarely between boats behind them and the finish line. Though the race will evolve in unforeseen ways, the crew will dutifully examine the wind maps and the current flows and adjust their plans.

I hope you can recognize the parallel with the market entry stage. Being in the leadership position is key to winning your race and similarly, you will have to jostle with your competitors even before the race starts. If you want to win your race, you need to study your market and know your competitive advantage. You can't just show up to the starting line and try a few strategies. You need to make a plan on how to start and *make decisions* that will win your entry market leadership, solidly differentiated from your competitors. This is the foundation of the Market Entry Strategy; I hope this analogy will speak to you now too.

I don't presume that this book and the associated methodology are a miracle approach, not any more than any regatta handbook can guarantee you a trophy. But I do hope it provides a useful framework for you to plan and make those decisions, as the feedback from my seminar attendees seemed to demonstrate and motivated me to organize my thoughts and write this book.

Acknowledgments

Like all books, this has been a labor of love and a much more daunting task than I imagined. I initially thought I could just take the seminar materials I'd created and turn them into a book, but I was wrong about that. Much help was needed to bring this book across the finish line.

The seminars were foundational to this book, and I first want to acknowledge the many people who encouraged me to develop the seminars, starting with Yair Vardi at the Fusion LA accelerator. Yair set the date for the first seminar and forced me to develop and transform the underlying materials into a class format. Many others have since hosted me as a seminar presenter at multiple locations as well as online. They helped bring the audience who welcomed the information, gave me support to continue in this endeavor, and assisted me to define and develop my ideas over the years.

The notion that I should write a book was formed over time as the thousands of seminar attendees praised the content but also asked for more details, more exercises, and to put my ideas into a format they could take with them as a guide through their initial entrepreneurial journey. The seminal push and approbation came from Mahendra Ramsinghani, author of several books but especially *The Business of Venture Capital*. Mahendra's book represented a level of quality, rigor, and insight that I could only aspire to. After he attended a seminar and gave me the initial thumbs up, I gained confidence that a book was the right format to go beyond the seminars.

As a foreign-born English speaker, I needed someone to take my raw materials and turn them into something that was digestible, as well as tame my overwhelming content production. Peter Birkeland was warmly recommended by Mahendra, Brad Feld, and Matt Blumberg, who had all written and published books with him on adjacent topics. Beyond his written contributions, Peter became a believer in the originality of the materials, an infinitely patient coach, and a critical support during the inevitable periods of self-doubt that could certainly have ended the project. He became a friend and the doula who brought this book to life.

When the book was in a somewhat readable format, a number of people whose opinions I greatly respect were kind enough to read and provide helpful suggestions for improvement: Gil Ben-Arzi, Nathalie Delrue-McGuire, Fara Hain, Steve Krausz, Erez Ofer, Casey Tansey, Yair Vardi, and Alon Yariv. I want to particularly highlight the feedback from Magdalena Yesil, a recognized entrepreneur and author of *Power Up: How Smart Women Win in the New Economy*, and Brian Fuller, formerly Editor-in-Chief and Silicon Valley bureau chief of EE Times. Both Magdalena and Brian went well beyond the assigned mission of "reviewing the manuscript" and became strong supporters of moving ahead and getting the book published. Finally, thank you to Karla Olson of BookStudio for publishing and marketing consultation; to Laurie Gibson, The Superior Editor, for her keen copyedit, to Lisa Wolff for her eagle-eye proofing, and to Claudine Mansour, Claudine Mansour Design, for a beautiful cover design and a clean and professional interior layout.

At this stage in my career, I have an infinite number of people to thank. Actually, each and every person that I met, sometimes directly in my professional endeavors, sometimes very far afield, in some way contributed to the knowledge and the life experience I tried to capture in this book. Particularly notable were my bosses, my collaborators, my board members and those I have sat on boards with, my USVP partners, and especially each and every entrepreneur I met, some for one hour, some for over a decade. There is something to be learned from each encounter. Some readers might recognize an anonymized version of what we experienced together. I hope I've related it decently and protected the innocent (and the not-so-innocent) but most importantly, that none of my writing and stories will be hurtful.

At the retirement dinner of a legendary entrepreneur a few years ago, the topic of legacy came up. For just about all of us, we won't be in the history books, and nobody will know of us in a few decades at most.

However, each of us gets to pass on an imprint of how we influenced people we knew and those people in turn have an opportunity to pass it on to others, sometimes not even knowing of the original influencer's existence. Many people have figuratively sat on my shoulder or whispered in my ear, and I hope this book will pass onward and pay forward the gifts they gave me.

About the Author

Dr. Jacques Benkoski has broad experience from academic and research organizations early in his career, to startups, scaleups, and finally venture capital during the past two decades. He has worked with companies with literally three guys in a garage and all the way through IPOs and M&As for both large and small firms. A technologist at heart, his career spans from the product and engineering functions of the semiconductor industry to consumer-oriented software. Since 2009 he has been a venture investor at U.S. Venture Partners (USVP) with a focus in cloud infrastructure, enterprise software, and cybersecurity. Jacques has held executive management positions, technical positions, and sales and marketing roles over the course of his career and has been directly responsible for bottom-line-generating revenue.

Jacques is passionate about helping entrepreneurs execute their vision and leverages his personal and professional networks to ensure that startups accelerate growth and build significant and sustainable businesses. His greatest satisfaction comes from seeing products he was involved with being used by customers and being able to help those entrepreneurs create and scale their business.

Earlier in his career, Jacques held research and management positions at IMEC in Leuven, Belgium, at IBM's Scientific Center in Haifa, Israel, and at STMicroelectronics in Grenoble, France. He is fluent in English, French, and Hebrew, and is an active global participant in the industries and topics in which he is involved. He speaks on the topics of international investing and IT and is regularly asked to participate on panels in the US and Israel. He strongly believes in paying forward to others by sharing his experience and insights through his Market Entry seminars.

Jacques holds a BSc in computer engineering from Technion-Israel Institute of Technology, and an MSc and PhD in computer engineering from Carnegie Mellon University. He has published more than thirty technical papers and holds one US patent. He lives in the San Francisco Bay Area and is an avid competitive sailor.